GENECOLOGY AND ECOGEOGRAPHIC RACES

Papers in the Biological Sciences
Presented at the 73rd Annual Meeting
of the Pacific Division AAAS on the
Occasion of the 100th Anniversary
of the Birth of Göte Turesson

Edited by

Arthur R. Kruckeberg
Richard B. Walker
Alan E. Leviton

PACIFIC DIVISION AAAS
San Francisco, California
1995

Library of Congress Cataloguing-in-Publication Data

American Association for the Advancement of Science. Pacific Division. Meeting (73rd : 1992 : University of California, Santa Barbara)
Genecology and Ecogeographic Races: Papers in the Biological Sciences / Presented at the 73rd Annual Meeting of the Pacific Division AAAS on the Occasion of the 100th Anniversary of the Birth of Göte Turesson ; edited by Arthur R. Kruckeberg, Richard B. Walker, Alan E. Leviton.
p. cm.
Includes bibliographical references and index.
ISBN 0-934394-10-5 (acid-free paper) : $28.95
1. Ecological genetics – Congresses. I. Turesson, Göte, 1892-1970. II. Kruckeberg, Arthur R. III. Walker, Richard B. IV. Leviton, Alan E. V. Title.
QH456.A47 1992
575.1 – dc20 95-9508
CIP

Pacific Division, American Association for the Advancement of Science
California Academy of Sciences, San Francisco, California 94118

Printed in the United States of America
First printing 1995

GENECOLOGY AND ECOGEOGRAPHIC RACES

GÖTE TURESSON
Photograph probably taken in the 1940s
Courtesy of the Hunt Institute for Botanical Documentation, Pittsburgh, PA

TABLE OF CONTENTS

List of Contributors and Participants for the Turesson Symposium (with addresses)

Andersson, Stefan
Department of Systematic Botany
University of Lund
Ö. Vallgatan 18-20
S-223 61, Lund, Sweden

Baker, Herbert G.
Department of Integrative Biology
University of California
Berkeley, California 94720

Bowman, Robert I.
Department of Biology
San Francisco State University
San Francisco, CA 94132

Broughton, David A.
Department of Zoology
University of Texas
Austin, Texas 78712

Chambers, Kenton L.
Department of Botany
Oregon State University
Corvallis, Oregon 97331

Conkle, Thomas
Pacific Southwest Forest and Range Experiment Station
P. O. Box 245
Berkeley, California 94701

Dunlap, Joan
College of Forest Resources
University of Washington
Seattle, Washington 98195

Kruckeberg, Arthur R.
Department of Botany , AJ-30
University of Washington
Seattle, Washington 98195

Mac Key, James
Department of Plant Breeding
Swedish University of Agricultural Sciences
S-750 07 Uppsala, Sweden

Lewis, Harlan
14280 Sunset Bvld
Pacific Palisades, California 90272

McMillan, Calvin
Department of Botany
Umiversity of Texas
Austin, Texas 78713

Millar, Constance
Pacific Southwest Forest and Range Experiment Station
P. O. Box 245
Berkeley. California 94701

Platenkamp, Gerrit A.
Jones and Stokes Associates, Inc.
2600 V Street
Sacramento, California 95815

Shapiro, Arthur M.
Department of Zoology
University of California
Davis, California 95616

Shaw, Ruth G.
Department of Ecology, Evolution and Behavior
University of Minnesota
St. Paul, Minnesota 55108

Singer, Michael
Department of Zoology
University of Texas
Austin, Texas 78712

Stebbins, G. Ledyard
Department of Genetics
University of California
Davis, California 95616

Svensson, Linus
Department of Systematic Botany
University of Lund
Ö. Vallgatan 18-20
S-223 61 Lund, Sweden

Thomas, Christian D.
School of Biological Sciences
University of Birmingham
Edgbaston, Birmingham, U. K.

Vasco, Daniel A.
Department of Zoology
University of Texas
Austin, Texas 78712

Walker, Richard B.
Department of Botany, AJ-30
University of Washington
Seattle, Washington 98195

White, Raymond R.
H. P. Harvey and Associates
P. O. Box 1180
Alviso, California 95002

Preface

Göte Turesson, born in Sweden in 1892, had a major impact on plant and animal biology through his development of genecology and of the ecotype concept. Although most of his work was done in Sweden, as a young man he earned B.S. and M.S. degrees in botany at the University of Washington. This led to interest in this region, first expressed by Professor Kenton Chambers, in commemorating the hundredth anniversary of Turesson's birth.

Thus a symposium "ECOGEOGRAPHIC RACES: From Turesson to the Present" was developed and held at the 73rd Annual Meeting of the Pacific Division, AAAS, on the campus of the University of California, Santa Barbara, in June, 1992. Most of the papers presented in the symposium were submitted for publication in the symposium proceedings, plus several more which were given there by abstract only. In addition, four biologists who have had long interest and activity in genecology submitted recollections or papers for the proceedings.

We believe that this volume, which includes all of the contributions mentioned above plus an introduction and a reprinting of a memorial written in 1971 by Turesson's friend and colleague, Professor Arne Müntzing, represents a fitting tribute to Turesson. In addition to historical aspects, the reader is given a broad picture of current applications of genecology in a range of plants and animals.

We thank the Pacific Division, AAAS for sponsorship of the symposium, and willingness to publish this volume of papers originating from it. It joins an outstanding group of volumes from previous symposia sponsored by the Pacific Division. Also we thank Michele Aldrich, liaison to the Pacific Division from the National Office, AAAS and chair of the Division's Committee on Publications, for her interest in and encouragement of the publication. Lastly, but assuredly not least, we thank those of our colleagues who agreed to review each of the contributions, and who then made many important and useful suggestions for their improvement.

The Editors

Ecogeographic Races: Turesson To The Present

Introduction

Arthur R. Kruckeberg
Department of Botany, AJ-30
University of Washington, Seattle, WA 98195

The noted Swedish botanist and geneticist, Göte Turesson, founder of genecology and the ecotype concept, received his first degrees in botany at the University of Washington (Walker, 1995, this volume). Also, his pioneering studies influenced the outstanding work on ecotypes of Clausen, Keck and Hiesey (1940) of the Carnegie Institution of Washington's laboratory based at Stanford, California. With this West Coast connection, it is fitting that the Pacific Division sponsor this symposium to commemorate the 100th anniversary of Turesson's birth.

The idea that plant species are not uniform and immutable, but possess both variability throughout their habitats and the capacity for change is not new. The first truly experimental studies of this variant nature of species began in the late 19th and early 20th centuries. Gaston Bonnier in Switzerland and Fredrick Clements in the United States first conceived of uniform garden tests for the inherited basis of species variants. Clements began varied environment studies on Pikes Peak, Colorado, about the time (ca. 1918) that Turesson was doing uniform garden experiments in Sweden. Harvey Monroe Hall joined Clements in 1918 as his active co-worker on the Colorado studies, thus preparing him for the later efforts in California (Clausen, Keck, & Hiesey, 1940).

It was Turesson, however, who performed the most extensive and controlled studies of species variation, pioneering the uniform garden technique in Sweden. From his papers beginning in 1922, Turesson initiated what would become a flourishing field of research: genecology, which has become the field of plant population biology in large part today. Indeed, the very concept was embodied in the title of his 1922 paper: "The genotypical response of the plant species to the habitat." And with the concept came a useful terminology, whose capstone term was the ecotype: "the product arising as a result of the genotypical response of an ecospecies to a particular habitat" (Turesson 1922).

The West Coast connection with Turesson's genecology was fixed firmly soon after his first publications. Harvey Monroe Hall (1874-1932) began in California the pioneering experiments to determine the nature of ecologically variable species (Hall 1926). It was Hall who invited Jens Clausen (1891-1969) to leave Denmark for a position in Hall's experimental taxonomy program. Following a year's (1927-28) stint at Stanford as an International Education Board fellow, Clausen became a staff member of Stanford's Carnegie group in 1931. Tragically, only four months later, Hall died; Clausen then was chosen to head the genecological program at Stanford (French 1989). The rest is familiar history to those nurtured in the exciting times of neo-Darwinian evolutionary biology, from the 1940s to the present.

It is fitting to look upon "the present" as a continuum from the recent past — Turesson to Clausen, Keck and Hiesey (1940) down to the present; though too numerous to single out, the figures in the field of population biology today, represented by such diverse talents as those of John Harper, G. L. Stebbins and Janis Antonovics. Contemporary population ecology and genetics are livelier than ever. What Turesson launched in the 1920s as the study of ecotypic variation in species thrives today as the study of infraspecific variation, its adaptive functions and its evolutionary potential. All major kingdoms of the biological world are suited to — and indeed profit from — the genecological approach. In this symposium, we witness examples of ecotypic variation in fungi, insects, and higher plants (both terrestrial and marine). Moreover the discipline's international flavor is exemplified in the Symposium by contributions from Sweden and North America. The invited papers that follow bear witness to the vitality of the Turesson legacy.

Literature Cited

Clausen, J., D. D. Keck, and W. M. Hiesey. 1940. Experimental Studies on the Nature of Species. I. Effect of Varied Environments on Western North American Plants. *Carnegie Inst. Washington*, Publ. 520, Washington, D.C.

French, C. S. 1980. Jens Christian Clausen 1891-1969. *Biographical Memoirs* 58:74-107. National Acad. Sci. Press, Washington, D.C.

Hall, H. M. 1926. The taxonomic treatment of units smaller than species. *Proc. Internat. Congr. Plant Sci.* 2:1461-1468. Ithaca, NY.

Turesson, G. 1922. The genotypical response of the plant species to the habitat. *Hereditas* 3:211-350.

Walker, R. B. 1995. Göte Turesson – The Pacific Coast connection. Pages 19-26 *in* A. R. Kruckeberg, R. B. Walker, & A. E. Leviton, eds., *Genecology and Ecogeographic Races*. Pacific Division, American Association for the Advancement of Science, San Francisco, CA.

HISTORY AND PERSONAL RECOLLECTIONS

Göte Turesson in Memorium

Arne Müntzing

(Reprinted with permission from *Bot. Notiser* 124:419-424, 1971)

Göte Wilhelm Turesson was born in Malmö in 1892 and went to school at first in Malmö and later on in the neighbouring town of Lund. As a schoolboy he was quite intelligent but rather obstinate and oppositional which led to conflicts with his teachers. Finally, when he had reached the age of 20, he stated with dismay that the Swedish school system was no good. Consequently, he emigrated to the United States, where he was received by his aunt who lived in Spokane, Washington, not far from Seattle.

After a year of hard struggle with various incidental employments in order to save money for his future activity Turesson matriculated at the State University of Washington in Seattle. Especially when considering his previous school problems in Sweden, his career at the American university was surprisingly rapid and successful. In 1914 he became a bachelor of science and one year later he acquired the master of science degree. He also got appointments as assistant, at first at the Department of Systematical Botany of the University, and later on at the Department of Physiological Botany.

In order to understand how this rapid development could be possible, it is necessary to realize that the childhood of young Göte was not only characterized by his oppositional tendencies but also by early positive influences of decisive importance. His father, Jöns Turesson, was a school-master with strong biological interests. Even before the age of ten, Göte made excursions with his father, collecting plants as well as butterflies, and quite early it was considered self-evident that Göte should become a botanist.

In 1915 he returned from America to Sweden, and in 1916 he matriculated at the University of Lund where he became a licentiate of botany in 1921. In the following year he defended his doctor's thesis and became a docent of botany. Before that he had temporary employments as assistant at the University Institutes of Systematical as well as Physiological Botany and also at the Institute of Genetics.

While still a schoolboy Göte Turesson published three small botanical articles, but his real scientific production was commenced in the United States with one ecological and one mycological paper. Another paper was a rather comprehensive plant-biological work, mainly concerning "skunk cabbage" (*Lysichiton camtschatcense*) and its role in the plant communities of bogs.

After his return to Sweden, Turesson at first continued his mycological work and wrote a comprehensive paper on the fungus flora in the intestinal tract of animals and humans. He also demonstrated that certain diseases in honey bees are caused by toxic substances produced by mould fungi. Turesson wrote a series of articles on bee diseases and became a well-known and often consulted bee-doctor.

By and by, however, this field of work had to give place to more central research projects with flowering plants. As early as in 1917 the genecological main line of his research work may be discerned in a paper on plagiotropism in sea shore plants. Turesson discovered that prostrate plant-forms comprised hereditary *prostrata*-forms as well as purely environmental modifications and that these two categories may be represented even within the same Linnean species.

1922 is an important year in Turesson's scientific production. In this year his doctoral thesis, "The genotypical response of the plant species to the habitat," was published and also two interesting forerunners. The first one deals with the development of the new science of plant sociology. In contrast to leading plant sociologists, Turesson realized that plant species are by no means uniform but are composed of races with different ecology. In the second forerunner-article, Turesson reported that the plant species are composed of different ecological units with different appearance and physiological properties. These so-called ecotypes are genotypically adapted to different habitats or climatic areas.

In the doctoral thesis all the empirical data, obtained from a large number of plant species, were accounted for. Living plants of these species had been collected from different habitats and were then cultivated and analysed in a garden belonging to the Institute of Genetics of the University of Lund.

Turesson realized that he had opened up a new and essential field of research, and in 1923 he published a declaration of program under the title "The scope and import of genecology." In this declaration he stressed that, so far, ecological research had been carried out without realization of the fact that there is a hereditary differentiation within species. Hence, the work initiated concerning the subdivision of species into genotypically different edaphic and climatic types represented a new phase in ecological research.

In numerous papers, approximately up to the middle of the nineteen thirties, Turesson presented the results of his large research program. This work was carried through with a remarkable strength and energy with regard to the collection of the large material as well as its detailed analysis in the experimental garden. Special attention was also devoted to the occurrence of characteristic physiological differences between the ecotypes — a work in part carried out at research institutes in Munich and Vienna.

The collection trips were at first limited to the province of Scania, South Sweden, but were by and by extended to all Sweden, and then to other countries and continents, including Siberia, United States and Canada.

In 1934, Turesson — as the sole European — was invited to attend the annual meeting of the Carnegie Institution of Washington which was this year held in Palo Alto, California. The specific research department of the Carnegie Institution located at Palo Alto was devoted to genecology, and the leader of this research group was the Danish-born biologist Jens Clausen. His research program was very much in line with Turesson's pioneer work and also included cytotaxonomy and experimental taxonomy.

Among the numerous publications following Turesson's journeys and expeditions a paper from 1925, "The plant species in relation to habitat and climate," is of special importance because it stresses the occurrence of parallel ecotypes in different species (*e.g.*, oecotypus *campestris, arenarius, salinus, subalpinus, alpinus*, etc.). Another paper, from 1930, "The selective effect of climate upon the plant species," contains other essential results concerning characteristic features in the genecological differentiation between widely separated flora regions (Atlantic versus continental, northern versus southern regions, etc.).

Besides the central concept ecotype, Turesson also introduced several other terms which chiefly concern the delimitation of different kinds of species from each other. In this connection he criticized the tendency of contemporary plant taxonomists to describe all plant forms that can be distinguished morphologically and are truebreeding as separate species. Especially among apomicts these criteria may lead to absurdities. Turesson, instead, proposed and defined the concept agamospecies which corresponds to natural species among plants with sexual reproduction.

For Göte Turesson the time from 1922 to 1927, when he was a docent of botany at the Lund University, was a very intensive and productive period that also involved personal happiness. In 1922, he married Benedicte Lehmann, and in 1924 their son Per Jöran was born. In 1927, Turesson left Lund and took up a position as chief plant breeder at the Weibullsholm Plant Breeding Institute in Landskrona. His breeding work with oats, rye and potatoes lasted four years but was abruptly terminated in 1931, when Turesson and three of his colleagues decided to leave their positions on account of a disagreement with the administrative leaders of Weibullsholm. Turesson then returned to the University of Lund where he got a position as "research docent." His new period of basic research work in genecology was, however, soon disturbed by a long and complicated period of competition for the chair of systematic botany in Lund. He was not appointed to this position, but in 1935 he became professor of systematical botany and genetics at the Agricultural College of Sweden in Uppsala.

During the first years in Uppsala, Turesson was much burdened by administrative duties — especially the construction and equipment of an institute and the selection of suitable co-workers. He was also much engaged in the establishment of a new genecological garden.

Concerning Turesson's research work during the Uppsala period, the following main features may be mentioned. In the nineteen forties he and his co-workers produced a series of polyploids in various cultivated plants. The best results were represented by tetraploid strains of red clover and alsike clover. Especially the latter one deserved attention since it probably represents the very first case of an experimentally induced autotetraploid of real economical importance.

Investigations on various apomicts had been started by Turesson during his period in Lund. The first papers of this kind concerned *Festuca ovina*, which comprises sexual races as well as more or less obligately viviparous forms. In this material, an autotetraploid series of chromosome numbers was detected, ranging from diploidy to hexaploidy. Turesson was also interested in the occurrence of apomictic microspecies in *Alchemilla*, and in papers during the period 1943 to 1958 he demonstrated that these microspecies are heterogenous and comprise many different biotypes. Another apomictic complex, investigated by Turesson and his wife, is the collective species *Hieracium pilosella*. This complex comprises sexual as well as more or less aposporous types with chromosome numbers ranging from 18 to 63.

Göte Turesson's last years in Uppsala were in several respects very burdening, and hence it was with relief that he reached the pension age in 1959. He was then happy to return to Lund and lived there 12 additional years. Though Turesson had not any more the great scientific force of his early manhood, a favourable effect of the return to his native province became obvious. Among other things, this new flourishing led to the comprehensive investigation of the *Hieracium pilosella* group.

Even as late as in the autumn of 1970 Turesson, together with some young botanists from Lund, made a collection trip to Jutland and had

Göte Turesson together with a student on an excursion to W. Jutland in September, 1970. Photo: Mats Gustafsson.

once more the pleasure of strolling around in the natural vegetation, making observations and collections. On account of a serious liver disease his strength was, however, clearly reduced and on the 30th of December, 1970, Göte Turesson died quietly in his home. During his last illness, he was quite aware that his time would soon be up, but he met the unavoidable without fear and with a scientific mind. He realized that life and death are only different aspects of the same basic biological phenomenon — a phenomenon that man shares with all other organisms in this world.

Göte Turesson was a very dynamic person whose manifestations of power and drastic mode of expression were sometimes very striking. With advancing age, his personality became progressively milder, and to his close friends it had all the time been clear that he possessed gentle and charming strings on which, however, he did not like to play in his earlier days. During his golden age, he used his oppositional mind and his great intellectual force for clearing new scientific paths, and indeed, this led to results of fundamental importance.

Arne Müntzing

Publications by Göte Turesson

(List prepared by Mrs. Madeleine Gustafsson)

1909 En jätteask i Skåne. *Fauna och Flora*, p. 99.

1910 Tjänstgör morkullan under sin flyttning såsom Goodyera repens fröspridare? Genmäle. *Ibid.*, p. 40-41.

1912 Några adventivväxter från Skåne. *Svensk Bot. Tidskr.* 6:95-96.

1914 Slope exposure as a factor in the distribution of *Pseudotsuga taxifolia* in arid parts of Washington. *Bull. Torrey Bot. Club* 41:337-345.

1915 *Penicillium avellaneum*, a new ascus-producing species. *Mycologia* 7(5):284- 287 (together with C. Thom).

1916 *Lysichiton camtschatcense* (L.) Scott, and its behavior in sphagnum bogs. *Amer. Journ. Bot.* 3:189-209.

1916 The presence and significance of moulds in the alimentary canal of man and higher animals. *Svensk Bot. Tidskr.* 10:1-27.

1917 The toxicity of moulds to the honeybee, and the cause of bee-paralysis. *Ibid.* 11:16-38.
1917 Om orsaken till binas s.k. Majsjuka. *Bitidningen* (Jan.), p. 11-14.
1917 Mykologiska Notiser. I. Ett fall av Aspergillusmykos hos bin. *Bot. Notiser*, p. 269-271.
1917 Om plagiotropi hos strandväxter. *Ibid.* p. 273-296.
1918 Binas sjukdomar. *In* A. Holm, *Handbok i biskötsel*, Ed. 2. p. 189-210.
1918 Om långväga växttrasport genom fåglar. *Bot. Notiser*, p. 248.
1919 The cause of plagiotropy in maritime shore plants. Contributions from the Plant Ecology Station, Hallands Väderö, No. 1. *Lunds Univ. Årsskr. N.F. Avd.* 2, 16(2):1-32.
1919 Grupp- och artbegränsning inom släktet *Atriplex*. *Bot. Notiser*, p. 41-47.
1919 Om utbredningen och bekämpande av Nosemasjukan. *Bitidningen* 18(7-8):145-147.
1920 Mykologiska Notiser. II. Fusarium viticola Thüm. infecting peas. *Bot. Notiser*, p. 113-125.
1921 Om olika slag av utsot hos bin. *Bitidningen* (Nov.), p. 233-234.
1922 Bisjukdomar under år 1921. *Ibid.* (Febr.), p. 24-26.
1922 Till frågan om bipestsjukdomarnas bekämpande. *Ibid.* (Oct.), p. 251-255.
1922 Växtsamhällslärans utveckling. *Bot. Notiser*, p. 49-68.
1922 The species and the variety as ecological units. *Hereditas* 3:100-113.
1922 The genotypical response of the plant species to the habitat. *Ibid.* 3:211-350.
1922 Über den Zusammenhang zwischen Oxydationsenzymen and Keimfähigkeit in verschiedenen Samenarten. *Bot. Notiser*, p. 323-335.
1923 The scope and import of genecology. *Hereditas* 4:171-176.
1924 Bisjukdomarnas uppträdande i vårt land och åtgärder för deras bekämpande. K. Landtbruksakad. *Handl. Tidskr.*, p. 1-15 (together with K.E. Sandberg and S.M. Tullberg).
1925 The plant species in relation to habitat and climate. *Hereditas* 6:147-236.

1925 Studies in the genus *Atriplex* I. *Lunds Univ. Årsskr. N.F. Avd.* 2, 21(4):1-15.
1926 Die Bedeutung der Rassenökologie für die Systematik und Geographie der Pflanzen. Feddes Repert. *Spec. Nov. Regni Veg. Beiheft* 41:15-37.
1926 Habitat and genotypic changes. A reply. *Hereditas* 8:157-160.
1926 Experimentell eller beskrivande växtsystematik? Ett inlägg. 13 pp. Lund.
1926 Studien über *Festuca ovina* L. I. *Hereditas* 8:161-206.
1927 Contributions to the genecology of glacial relics. *Ibid.* 9:81-101.
1927 Untersuchungen über Grenzplasmolyse- und Saugkraftwerte in verschiedenen Ökotypen derselben Art. *Jahrb. Wiss. Bot.* 66(4):723-747.
1927-1931 A number of brief reports about breeding resuts at Weibullsholm Plant Breeding Institute, Landskrona. Weibulls Årsbok (together with H. Erickson).
1928 Castration experiments in *Hieracium umbellatum* L. and *Leontodon autumnalis* L. *Svensk Bot. Tidskr.* 22:256-260.
1928 Erbliche Transpirationsdifferenzen zwischen Ökotypen derselben Pflanzenart. *Hereditas* 11:193-206.
1929 Zur Natur und Begrenzung der Arteinheiten. *Ibid.* 12:323-334.
1929 Ecotypical selection in Siberian *Dactylis glomerata* L. *Ibid.* 12:335-351.
1930 Studien über *Festuca ovina* L. II. Chromosomenzahl und Viviparie. *Ibid.* 13:177-184.
1930 The selective effect of climate upon the plant species. *Ibid.* 14:99-152.
1930 Zur Frage nach der Spontanität von *Betonica officinalis* L. in Schweden. *Bot. Notiser*, p. 495-506.
1930 Genecological units and their classificatory value. *Svensk Bot. Tidskr.* 24:511-518.
1931 Studien über *Festuca ovina* L. III. Weitere Beiträge zur Kenntnis der Chromosomenzahlen viviparer Formen. *Hereditas* 15:13-16.
1931 Über verschiedene Chromosomenzahlen in *Allium schoenoprasum* L. *Bot. Notiser*, p. 15-20.

1931 Field studies and experimental methods in taxonomy. *Hereditas* 15:1-12 (together with A. Müntzing and O. Tedin).
1931 The geographical distribution of the alpine ecotype of some Eurasiastic plants. *Ibid.* 15:329-346.
1932 *Trapa natans* L. im Altai-Gebiet. *Bot. Notiser*, p. 177-190.
1932Die Genenzentrumtheorie und das Entwicklungszentrum der Pflanzenart. *Ibid.* 2(6):1-11.
1932 Die Pflanzenart als Klimaindikator. *K. Fysiogr. Sällsk. Lund Förh.* 2(4):1-35.
1933 Zur Rassenökologie von *Adonis vernalis* L. *Bot. Notiser*, p. 293-304.
1933 Bemötande av några punkter i professor N. Heribert Nilssons "Sic et non." 9 pp. Lund.
1933 Professuren i systematisk botanik vid Lunds universitet. Ett inlägg. 25 pp. Lund.
1933 Återbesättandet av den lediga professuren i systematik botanik vid Lunds universitet. Besvär och förklaring. 17 pp. Lund.
1933 Ille faciet. Till belysning av professor Nordhagens sakkunnigutlåtande vid återbesättandet av professuren i växtbiologi i Uppsala. 16 pp. Lund.
1935 Växtartens ekologiska differentiering med särskild hänsyn till kulturväxterna. *Nord. Jordbrugsforsk.* 4.-7. Hefte, p. 547-552.
1936 Rassenökologie und Pflanzengeographie. Einige kritische Bemerkungen. *Bot. Notiser*, p. 420-437.
1938 Chromosome stability in Linnean species. *Ann. Agr. Coll. Sweden* 5:405-416.
1939 North American types of *Achillea millefolium* L. *Bot. Notiser*, p. 813-816.
1940 Rasekologiska problem. Vetenskap av i dag. p. 485-512.
1943 Variation in the apomictic microspecies of *Alchemilla vulgaris* L. *Bot Notisr*, p. 413-427.
1943 Chromosome doubling and cross combinations in some Cruciferous plants. *Ann. Agr. Coll. Sweden* 11:201-206 (together with H. Nordenskiöld).
1946 Kromosomfördubbling och växtförädling. *Weibulls Årsbok*, p. 1-8.
1949 Institute of Plant Systematics and Genetics. The Royal Agricultural College of Sweden. *Ann. Agr. Coll. Sweden* 16:75-77.
1956 Variation in the apomictic microspecies of *Alchemilla vul-*

garis L. II. Progeny tests in agamotypes with regard to morphological characters. *Bot. Notiser* 109:400- 404.

1957 Variation in the apomictic microspecies of *Alchemilla vulgaris* L. III. Geographical distribution and chromosome number. *Ibid.* 110:413-422.

1958 Observations on some clones and clone progenies in *Alchemilla alpina* L. *Ibid.* 111:159-164.

1960 Experimental studies in *Hieracium pilosella* L. I. Reproduction, chromosome number and distribution. *Hereditas* 46:717-736 (together with B. Turesson).

1961 Habitat modifications in some widespread plant species. *Bot. Notiser* 114:435- 452.

1962 Results of colchicine doubling in the red, alsike and white clover. *Agri Hort Genet.* 20:111-135.

1963 Observations on chromosome number and reproduction in some Piloselloids. *Bot. Notiser* 116:157-160 (together with B. Turesson).

1963 *Sedum anglicum* Huds. funnen på Christiansö. *Ibid.* 116:105-108.

1964 *Sedum anglicum* Huds. på Christiansö. *Ibid.* 117:426-427.

1966 Genecological notes on *Allium schoenoprasum*. *Trans. Bot. Soc. Edinb.* 40(2):181-184.

Göte Turesson: The Pacific Coast Connection

Richard B. Walker

Botany Department, AJ-30
University of Washington, Seattle 98195

Turesson was born in 1892 in Malmö, Sweden, and received his early education there. At age twenty he came to Spokane, Washington, to the home of an aunt. While working for a year at various jobs, and not yet educated as a botanist, he nonetheless started observations of the effects of slope and aspect on the distribution of Douglas-fir. The next year he enrolled at the University of Washington, Seattle, and came under the influence and tutelage of the three faculty members of the Department of Botany: Theodore C. Frye, morphologist; John W. Hotson, mycologist; and George B. Rigg, physiologist and ecologist. Along with earning the B.S. in 1914 and the M.S. in 1915, he finished and published the work on Douglas-fir (*Bull. Torrey Bot. Club* 41:337-345, 1914), reported with C. Thom a new species of *Penicillium* (*Mycologia* 7:281-287, 1915), and carried out a detailed ecological study of the western skunk cabbage (*Amer. Jour. Bot.* 3:189-209, 1916). Thus when he returned to Sweden in 1915 and matriculated at the University of Lund for doctoral studies, he was already an active botanist with three substantial publications to his credit. He came to the United states twice again: in 1934, when invited to attend the annual meeting of the Carnegie Institution of Washington, held that year in Stanford, California; and in 1937 for extensive travels in the field.

Other contributions to this symposium will bring out the influence of Turesson's scientific contributions on the biological sciences, but in this paper I will concentrate on him as a young student and scientist in Washington State. For this the records are scanty, but the University transcript, his papers written in the Northwest, and some statements made by his mentors are available. Also, his friend and colleague in Lund, Professor Arne Müntzing, wrote a memorial biographic tribute to Turesson (*Bot. Notiser* 124:419-424, 1971; reprinted, 1995, this volume, pp. 7-18), which has some information about the early years in America.

TURESSON AS A STUDENT

Today the graduate of a European gymnasium still has had training well beyond that of the American high school. Nonetheless, it is a little surprising that he was accorded junior standing with 90 credits upon entrance to the University of Washington on the recommendation of Professors Frye and Vickner, with the approval of Dean Landis of the College of Science. Further, the transcript states that he would require only 30 more semester credits to graduate with the B.S. degree. He passed the senior examination on June 8, 1914. Table 1 details the course work which he took in his two semesters as an undergraduate student (1913-14) and the two semesters as a graduate student (1914-15). In addition, in the summer of 1915, he was enrolled at the Friday Harbor field station. In the course work he received mostly A grades, a few B's, but D's in chemistry — many a biologist has stumbled a little in that subject. Clearly he had a concentrated course of study in the sciences and was awarded the B.S. in 1914 and the M.S. in 1915.

TURESSON'S MENTORS AT THE UNIVERSITY OF WASHINGTON

When Turesson came to the University of Washington in 1913, the faculty in the Department of Botany consisted of three vigorous teachers and researchers in their early forties. T. C. Frye had taught five years in high schools and a year in a small college before finishing his Ph.D. in plant morphology in the renowned Botany Department of the University of Chicago, then came to Seattle to found the Botany Department of the University of Washington in 1903. George B. Rigg had taught over ten years in high schools and a teachers college before taking a M.S. at the University of Washington and joining the faculty as an instructor, while working for the Ph.D. degree at the University of Chicago in plant physiology. When Turesson arrived, Rigg had just finished the Ph.D. and had been promoted to Assistant Professor. John W. Hotson was still working on the Ph.D. in mycology at Harvard University when he joined the faculty as an instructor in 1911. He finished the degree in 1913 but was not promoted to Assistant Professor until the next year.

Frye was an imposing square-shouldered man over six feet tall, and Rigg was almost as tall but more slight of build, while Hotson was shorter and more stocky. Figure 1 depicts the three as they appeared some fifteen years after Turesson was in the Department. In the 41 semester credits which he earned in Botany, he had ample contact with these three professors, each of whom was well-trained, able in teaching, and also research oriented. Also Professors Frye and Rigg always taught at the Friday Harbor field station, where Turesson was enrolled in the summer of 1915 just as he was finishing his master's degree. Although he received grades in the General Botany for Forestry Students, I think that he was serving as Dr. Hotson's assistant in these courses. Indeed, Turesson is listed as an assistant in Botany in the 1914-15 University catalogue, and Professor Müntzing noted that he served as an assistant in both systematic and physiological botany. He also fulfilled a minor in Zoology, taking a course in General Physiology and two courses in Entomology. The instructor in the latter was Professor Trevor Kincaid, an active, versatile biologist.

TABLE 1. Courses taken by Göte Turesson as a Student at the University of Washington.

Year and Semester	*Course No.*	*Subject*	*Credits*	*Grade*	*Instructor*
1913-14					
Autumn	Bot. 41	General Fungi	4	A	Hotson
	Bot. 5	Thallophytes	4	A	Frye
	Engl. 1	Grammar/Composition	4	B	
	Zool. 11	General Physiology	4	B	Smith
Spring	Bot. 6	Bryophytes/Pteridophytes	4	B	Frye
	Bot. 20	Plant Histology	3	C	Frye
	Bot. 42	Plant Pathology	4	A	Hotson
	Engl. 2	Grammar/Composition	4	B	
	Chem. 1b	General Chemistry	4	D	
Summer	Chem. 11	General Chemistry	2	D	
	Chem. 11a	Chemistry Laboratory	2	D	
1914-15					
Autumn	Bot. 11[1]	General/Foresters	4	A	Hotson
	Bot. 43	Physiological Processes	4	B	Rigg
	Bot. 50	Algae	2	A	Frye
	Zool. 17	General Entomology	4	A	Kincaid
Spring	Bot. 11[1]	General/Foresters	4	A	Hotson
	Bot. 44	Growth/Movements	4	A	Rigg
	Bot. 54	Angiosperms	4	B	Frye
	Zool. 18	Entomology	4	A	Kincaid
Summer	Bot.	Friday Harbor	1		

[1] Probably he was a teaching assistant in Bot. 11.

FIGURE 1. Professors J. W. Hotson, T. C. Frye, and G. B. Rigg, who taught Turesson in the Department of Botany, University of Washington.

TURESSON THE YOUNG SCIENTIST

Professor Müntzing informs us in his memorial biographical sketch that Turesson's father was a school-master with strong biological interests and that young Göte went on excursions with him, and collected plants. In fact from these collections he published three small taxonomic papers in Swedish periodicals. Thus it is not surprising that he started botanical observations around Spokane when he came there to live with an aunt in 1912. Most of the study for the paper "Slope exposure as a factor in the distribution of *Pseudotsuga taxifolia* in arid parts of Washington" (*Bull. Torrey Bot. Club* 41:337-345, 1914) was obviously done in the vicinity of Spokane (Fig. 2), although he also used some observations made in the San Juan Islands, indicating that he probably visited the Friday Harbor field station in 1913 or 1914.

The work in mycology must certainly have been done under the tutelage of Professor Hotson. In 1913-14 he had studied both general mycology and plant pathology under Hotson. Presumably while studying mycology, and perhaps as a project in the class, he started

Higher up on the steep slope, where *Pseudotsuga* becomes dominant, the vegetation shows the following composition:

TREES

Pseudotsuga taxifolia

SHRUBS

Ceanothus sanguineus | Schizonotus discolor
Opulaster pauciflorus | Spiraea corymbosa
Sambucus glauca

HERBS

Antennaria rosea | Geranium viscossisimum
Arnica cordifolia | Heuchera cylindrica

FIGURE 2. Plants which Turesson listed in his paper on Douglas-fir (*Pseudotsuga*).

the cultures from the bear dropping (Fig. 3) in which he isolated a sporulating *Penicillium.* At that time Dr. Charles Thom of the U.S. Department of Agriculture was the foremost authority on the *Penicilliums*, so presumably he sent the interesting form to Dr. Thom, who confirmed it as a new species, which they reported in *Mycologia* (vol. 7:281-287, 1915).

The study of *Lysichiton camtschatcense,* the western skunk cabbage (Fig. 4), was most likely a master's degree project, and presumably done with Dr. Rigg's supervision, since Rigg was a specialist in peat bogs. In any case, it was a detailed and interesting study which appeared in print (*Amer. Jour. Bot.* 3:189-209, 1916) after Turesson had already returned to Sweden.

There is no question that Turesson had a "running start" in botany when he came to the University of Washington, but that he vigorously pursued course work, field studies, and publication of his results while here. It is tantalizing to speculate as to whether his observations of plants in the Pacific Northwest might have led to any thinking about variations within species. There is nothing in the three papers from his work here to indicate such. However, Dr. Frye often taught a course (not taken by Turesson) entitled "Ecology: the factors causing environmental adaptations in plants," and one might expect that they had some discussions of such matters.

Certainly Turesson made the most of the opportunities here in the Northwest and at the University of Washington, and was obviously a vigorous professional botanist by the time he returned to Sweden late in 1915.

PENICILLIUM AVELLANEUM, A NEW ASCUS-PRODUCING SPECIES[1]

CHARLES THOM AND G. W. TURESSON

Ascus production by species of *Penicillium* is not common. The observations of certain species by Brefeld,[2] Morini[3] and Westling[4] have never been repeated by other workers, some of whom have watched thousands of cultures in the hope of finding one of these forms. On the other hand, *P. luteum* Zukal[5] is a member of a widely distributed group,[6] some members of which have been found repeatedly, while the ascus-producing form is not uncommon. Ascus production in this species is not dependent upon special methods of culture. Another species has now been found by one of us (Turesson) in cultures from the faeces of a bear in the Zoological Garden, at Seattle, Washington. In this form as in *P. luteum,* the asci are produced in almost all of the media regularly used. The time required varies from six weeks to perhaps three months. Its morphology relates it to *P. luteum* and to the ascus-producing forms of *Aspergillus*.

FIGURE 3. Excerpt from the paper by Thom and Turesson reporting on a new species of *Penicillium.*

FIGURE 4. Picture of skunk cabbage (*Lysichiton camtschatcense*) taken from Turesson's ecological paper on this species. Author's caption reads, "Fig. I. *Lysichiton camtschatcense* at time of flowering. Author's photograph."

CONNECTIONS WITH THE CARNEGIE GROUP AT STANFORD

Beginning about 1926 the laboratory of the Carnegie Institution of Washington at Stanford, California, developed a major program for the study of ecotypic variation in plants, especially as affected by altitude. This group was led from 1931 by Dr. Jens Clausen, a native of Denmark, with colleagues Drs. David Keck and William Hiesey. They were, of course, well acquainted with Turesson's pioneering work in genecology, and Clausen very likely knew Turesson personally. In any case, when the annual meeting of the Carnegie Institution in 1934 was to be held at the Stanford laboratory, Turesson was invited to attend and did participate. The frontispiece depicts him as a vigorous man of middle age, much as he must have looked in 1934, although this picture probably was taken a few years later.

Now you know why we consider Göte Turesson "one of our own" here on the Pacific Coast. As the papers to follow will show, the influence of his work has remained very much alive over the years since the seminal paper from his doctoral thesis was published ("The genotypical response of the plant species to the habitat," *Hereditas* 3:211-350, 1922). As Professor Müntzing points out, administrative and other burdens interfered some with Turesson's research in the middle years, but he continued scientific work after retirement. A photograph in Müntzing's paper (reprinted in this volume; see page 13) shows him in the field with a student only about three months before he died at age seventy-eight.

Recollections of Göte Turesson, His Garden and Collection Sites

Harlan Lewis
Department of Biology
University of California, Los Angeles

I first became aware of the work of Göte Turesson through a course I took at UCLA from Carl Epling about 1939. At the same time I also learned of the work of the Carnegie Group: Clausen, Keck and Hiesey, at Stanford. As a result, experimental taxonomy became my principal interest and I immediately applied the approach of growing material of the same species from different climatic regimes to the group of plants I was studying. At the time I was working on a taxonomic revision of the genus *Trichostema* using standard herbarium techniques, a study that had been suggested by Carl Epling, who was interested in completing a taxonomic treatment of the Labiatae for the North American Flora. Herbarium collections indicated that material of a then undescribed species growing at an elevation of about 2500 m in the San Jacinto Mountains of southern California was a dwarf, compact plant compared with populations of apparently the same species growing at elevations of 1000-1500 m in the same mountains. Consequently, I collected seeds of these annuals and grew progenies under uniform cultural conditions at the University of California, Los Angeles. The populations proved to be conspecific and the high elevation population retained its dwarf, compact habit in cultivation. I was elated to have demonstrated the existence of a distinct ecotype that warranted, in my estimation, subspecific recognition.

An opportunity to meet Göte Turesson came when my wife, Margaret, and I attended the Eighth International Congress of Genetics in Sweden in the summer of 1948. By this time I had an appreciation of the influence Turesson had had on experimental taxonomy and was familiar with exciting work that had been going on in California and elsewhere.

The previous spring, for example, I had a memorable visit with J.

W. Gregor at the Scottish Plant Breeding Station in Corstorphine near Edinburgh where we discussed his material of *Plantago maritima* and clinal ecotypic differentiation along various gradients.

Attending the Congress in Sweden was an exciting experience for one just starting his professional career. It was relatively small as such meetings go and provided an opportunity not only to hear but also to actually meet many people who had been only names in textbooks or on publications before. I was disappointed that Turesson was not scheduled to present a paper during the formal sessions in Stockholm. However, during the Congress one day was devoted to an excursion to Uppsala where one of the options was a visit to Turesson's experimental garden. At the garden Turesson discussed its establishment about a decade earlier as a successor to his garden in Åkarp where the classical work published in the 1920s was conducted. The setting of the garden near Uppsala was delightful and rather park-like, as I recall, suggesting to me that this was more of a demonstration than a working garden. Nevertheless, it was a thrill to follow Turesson through the garden as he pointed out some of the classic ecotypes of his early studies which he still maintained in this newer garden. I remember being particularly impressed with some of the conspicuously different edaphic ecotypes of *Hieracium umbellatum* and a rather bewildering array of variation in *Alchemilla vulgaris* which did not fall into neat ecotypic patterns. There were a number of plantings of trees and shrubs that presumably showed ecotypic differences, but if so they made no lasting impression on me.

An unscheduled, informal meeting was arranged by someone to meet with Turesson to discuss his concept of ecotype. I do not recall exactly when or where it was held nor even how I happened to be included. The group who met did not question the significance or prevalence of ecotypic variation but was interested in discussing the objectivity of ecotypes and their relation to ecogeographic variation recognized taxonomically as subspecies. Turesson, vigorously supported by Jens Clausen, maintained that ecotypes are objective, genetically definable entities. Because ecotypes by definition are interfertile, hybrid intermediates are found where ecotypes come in contact, but these intermediates, argued Turesson and Clausen, in no way negated the objective nature of ecotypes. Others argued that although sharply differentiated ecotypes may occur, especially where they are separated by abrupt edaphic features, the more general pattern of ecotypic differentiation was gradual and clinal; morphologically recognizable, genetically distinct ecotypes that warranted formal recognition were the exception. I made no contribution to the discussion. No consensus was reached, as I recall, and one of my impressions was that Turesson's major contributions were probably behind him. Nevertheless, just listening to Turesson express and defend his position was a great experience for me.

A delightful sequel to visiting Turesson's garden came in the summer of 1960 when Borje Lövkvist, who had often been in the field with Turesson, invited me to join him on an excursion to visit many of the sites in southern Sweden where Turesson had collected material for his classic ecotype studies. We traveled from Lund around the southern end of Sweden and up the east coast to north of Skania where we crossed to the west coast and eventually returned to Lund. Our stops were mostly near the coast because a great many of the early samples studied by Turesson were from various coastal habitats in southern Sweden. I do not remember how many of the numerous species studied by Turesson we encountered, but some of the sites and several of the species were particularly memorable. Early one morning, for example, we took a small ferry to Hallands Väderö, which according to Lövkvist was one of Turesson's favorite destinations. Hallands Väderö is a small island about 5km off the northwest coast of Skania where we spent the entire day exploring the various habitats and visited some of the precise sites of several of his collections. Although the island is of rather low relief, it includes several very distinct habitats ranging from exposed coastal rocks, salt and freshwater marshes, swampy areas to shady beech forest in the interior. In the interior I recall observing a shade form of *Lysimachia vulgaris* which Turesson had shown was not a shade ecotype but merely a modification. I also remember examining populations of *Sedum maximum* which were abundant not only on the rocky shores but also in the interior where it appeared quite different. We had seen similar differences at various sites on the mainland where Turesson also had collected. I recall being shown specific sites of collections of *Armeria vulgaris* and *Hieracium umbellatum.* The latter was particularly interesting because I remembered it from Turesson's garden and also because it was one of the most conspicuous plants in flower at most of the lovely coastal sites we visited. I remember being particularly impressed with the ecotypic differentiation between the rocky shore and dune populations on the mainland opposite Hallands Väderö, where one could step from a population of the rocky ecotype to morphologically quite different plants growing on the dunes. I have little doubt that such striking differences between adjacent edaphic ecotypes were responsible for Turesson's concept of ecotypes as objectively definable entities.

Although my encounters with Göte Turesson, his garden and collecting sites were brief, they gave me an appreciation of the man and the significance of his work, particularly as it has influenced my own understanding of ecotypic differentiation.

Göte Turesson: A Pioneer of Plant Experimental Taxonomy

G. Ledyard Stebbins
University of California, Davis

The intense research career of Turesson from 1922 to 1932 heralded a turning point in the history of plant taxonomy, particularly its relationships with other disciplines of botany and of evolutionary biology in general. In all of biology, this decade was marked by a decline of parochialism and a rise in international communication between scientists. It was also marked by a realization that in all of biology, including systematic botany, the results of experiments, carefully designed and conducted with meticulous care to detail, were necessary before answers to the big basic questions could be intelligently considered. For answering the biggest question that in 1922 was still unanswered: How much of the differences between species and individuals, as we see them in nature or on dried specimens, is based upon the plant's genes and how much is due to environmental modification during development?, the state of knowledge in 1920 was most favorable to progress. Much, and perhaps all of heredity was known to be based upon particulate genes located on chromosomes and transmitted via eggs and pollen or sperm; experiments being conducted by E. M. East were showing that so-called quantitative inheritance could be based upon large numbers of particulate genes; and those of W. Johannsen on pure lines of beans had demonstrated that individual differences based upon differences in environment were not inherited. One could logically reason, as Turesson did, that experiments on plants that were derived directly from nature would provide answers that would be valid for these plants and the species to which they belonged. When, therefore, Turesson's three seminal papers appeared in 1922-1923, many taxonomists realized at once that here was the wave of the future in their discipline. The most enthusiastic of them, J. C. Clausen, J. W. Gregor, H. M. Hall, E. N. Sinskaia and W. B. Turrill, discussed actively both Turesson's results and their own, in relation to the ever controversial species problem. Turesson's influence by 1940 had come to dominate the theoretical basis of plant systematics, as regarded by ecologists and evolutionists.

His research and ideas, important as they were, must be balanced against his theoretical shortcoming, which was shared by some of his followers. The chief shortcoming was adherence to a type concept. To him, each ecotype was a distinct entity, described by the characteristics of the plants that he grew and observed and forming a pattern of discontinuous variation with respect to other ecotypes that belonged to the same species. This condition was at least partly true for some of his edaphic ecotypes, like those adapted to sandy seashores and coastal bluffs, but his geographic ecotypes were in large part an artefact of sampling, since they were based upon a small number of localities for a widely distributed species. Another Swedish botanist, Olof Langlet, showed that in the dominant tree species of Sweden, *Pinus silvestris,* patterns of geographic variation were mostly continuous from one population to another, and discontinuities always reflected discontinuities in climatic or edaphic factors. Similar patterns were found by Gregor in the *Plantago maritima* complex and by various other workers. Because of these facts, most taxonomists have discarded the term ecotype, and refer to infra-specific entities as subspecies or races.

Although Turesson was active for many years after 1932, he was beset by administrative duties and health problems that caused a marked decline in productivity. His most significant later publications dealt with experiments on asexually reproducing plants; *Festuca, Alchemilla* and *Hieracium.* These papers were far less innovative than his earlier work and, compared with contributions of others to this field, were of minor importance.

During his most active period, Turesson's personality was like his research: dynamic, enthusiastic and unstoppable. Jens Clausen, one of his greatest admirers, loved to repeat an anecdotal event that took place during his first visit to California in 1934, when Clausen had just begun his classic experiments on experimental taxonomy and asked Turesson to stay at his home during an important conference. Following a custom for his visitors, Clausen was driving Turesson along the highway west of Palo Alto, in order to show him the native redwoods in the Santa Cruz mountains. Shortly after they had left the town, Turesson looked out of the car window at a patch of live oak scrub and asked: "Is that bush poison oak?" "Yes." "Stop the car, I want to get out!" Clausen dutifully did so, Turesson jumped out, plunged into the scrub, took off his shirt, picked poison oak leaves, and rubbed his chest with them, saying: "Now I'll know for sure whether I'm susceptible or immune!" Fortunately, he was immune.

I met Turesson on two occasions. In 1937, he visited California on a tour of the western states, organized and guided by students from the University of Minnesota. Our group, consisting of E. B. Babcock, J. Clausen, W. M. Hiesey and myself, joined his party at Mather, where we had an extended discussion of the plantings of *Potentilla, Achillea* and other genera in the middle altitude garden. After we had spent the night at Mather, our caravan continued to Tuolumne Meadows and the Timberline Station, where a review of the high altitude ecotypes was enriched by a discussion of the subalpine flora compared to similar floras that Turesson knew in Scandinavia. Then down to Leevining and a restaurant on the shore of Mono Lake, where beer flowed freely and Turesson, completely relaxed, was his energetic and entertaining self. During this part of the trip we got an interesting account of Turesson's entire journey, from the students who were his guides. Although they were thrilled to be with him, and profited much from what he had to say, his personal habits were somewhat disturbing to them. These young people, brought up in country homes of Minnesota, were used to early rising, frugal habits, and early to bed. Turesson, on the other hand, was in his element in after dinner talk washed down with plenty of beer, so that midnight bedtime, and rising after eight in the morning, was normal for him. They were having the time of their lives, but not without cost to their normal habits.

The second time that I met Turesson was after the Eighth International Congress of Genetics held at Stockholm in 1948. Turesson did not present a paper at the Congress, but after it many geneticists visited Uppsala, where he was a Professor, and he gave us a tour of his garden. At that time, it contained chiefly apomictic clones of the genus *Alchemilla*, among which he had isolated several mutants and had evidence of chromosomal segregation. He was no longer the ebullient personality of former years, having suffered a setback in not receiving the appointment at Lund that he intensely desired and being at Uppsala overburdened with administrative duties. Regretfully, we realized that at the age of 57, he had virtually completed his scientific career. When I returned to Sweden in 1961, Turesson was retired, living at his home in Lund, and not anxious to see foreign visitors.

Even though my contacts with Göte Turesson were all too brief, my impression of him as a dynamic pioneer in our field is still vivid and rewarding.

GENECOLOGY
SYSTEMATICS
POPULATION BIOLOGY

The Contributions of Göte Turesson to Plant Taxonomy

Kenton L. Chambers
Department of Botany & Plant Pathology
Oregon State University, Corvallis, OR 97331

Göte Turesson's contributions to taxonomy, especially when placed in a historical perspective, were both significant and long lasting. During a brief span of time in the 1920s and early 1930s, when his theories of genecology were having their initial resounding impact, he earned a position as founding father of a new field of botanical research known alternatively as biosystematics or experimental taxonomy. The development and blossoming of this branch of taxonomy has been reviewed by previous writers (Hagen, 1983, 1984) and is too well known a story to require retelling in its entirety. This seems especially true now as we near the end of the 20th century and note how radically systematics has changed since Turesson's time. Recent rapid advances in such areas as population genetics, molecular phylogenetics, computer-assisted cladistics and phenetics, and chemotaxonomy have revolutionized systematics. Nonetheless, some selective comments about Turesson's contributions, made from the perspective of a biosystematist whose career began during the "glory days" of that field, may be a useful contribution towards the yet-to-be-written comprehensive history of plant taxonomy in the modern era.

THE CONDITION OF TAXONOMY A CENTURY AGO

To many present-day workers, Turesson must seem to be a figure from the distant past, and indeed we are now farther removed in time from him than he was, in turn, from Charles Darwin. In order to appreciate his contribution to the modernizing of plant taxonomy, we must review some of the problems faced by practitioners of systematics early in the present century. Probably more than at any time before or since, philosophical disagreements among taxonomists were extremely marked. For most biologists, a static, pre-Darwinian view of species had been replaced by a concept in which variation and evolutionary change were Nature's rule. The division between taxonomic splitters and lumpers was especially prominent, with a few of the more extreme splitters still operating under a creationist view of the origin of species. Even among those who advocated evolutionary principles, there arose different "schools" of workers divided somewhat along the lines of traditionalists vs. reformers. Philosophical disagreements, both in the realms of taxonomy and nomenclature, were in the United States, for example, connected with major institutions such as Harvard University and the New York Botanical Garden, which were dominant and competing players in turn-of-the-century American botany. A tradition of broadly circumscribed genera and species, the legacy of Asa Gray, dominated at Harvard, while members of the New York "school" often tended to fragment established genera and to describe new species at the slightest provocation (see, for example, the treatment of *Astragalus* by P. A. Rydberg, 1929).

Primarily at fault in this period of malaise in American taxonomy was the persistence of typological species concepts (Mayr, 1987) held over from the time when external morphology was the only source of data for plant systematics. Yet to come were the integrative, ecogenetic species concepts of Turesson and the "experimental taxonomists," to whom each species was the sum total of its constituent ecological and geographical races, no one of which was "typical" in other than a nomenclatural sense. Because of the typological approach used by many taxonomists, classifications became highly subjective, lacking in common principles and varying widely from worker to worker. A legacy of that era is the load of generic and specific synonyms which encumbers the nomenclature of so many plant groups. For most taxonomists of the time, figures are not available on scientific "batting averages" — that is, the percentages of their described new species and varieties that are recognized in the current floristic and monographic literature. However, the biography of Aven Nelson (Williams, 1984) is unique in listing all the taxa described by that botanist along with their present status as accepted names or synonyms. Nelson is widely admired for his pioneering work on the flora of the Rocky Mountains; his publications spanned the pre- and post-Turesson era (1896-1945). Nonetheless, the success rate of his taxonomic proposals was modest, to say the least; 75% of his 796 new taxa (694 species and 102 varieties) have been reduced to synonymy (data from Williams, 1984).

The fact that systematics has a strong subjective component, and that one person's "species" are another's "varieties" or even "minor genetic variants," is well understood by practicing taxonomists. In the historical period under discussion, however, when the intellectual ferment of Darwinian evolution had been at work for some 40 years, it must have been frustrating to many biologists that taxonomy had not updated its concepts and procedures to reflect an evolutionary view of species variability. Worse than the bickering between rival "schools" was the image of botanical systematics as a field mired in the past and plagued by uncritical research and incommensurable results.

THE EXAMPLE OF EDWARD L. GREENE

It may appear unseemly to complain about the scientific errors and inadequacies of earlier workers in one's field. Yet the work of Edward L. Greene, whose taxonomic career spanned the period between Darwin and Turesson, offers a useful perspective on the then brewing revolution in plant systematics. Through Greene's adoption of a narrow, typological species concept, his voluminous publications added weighty lists of synonyms to the names of western North American plants. In the annual section of *Microseris* (Asteraceae), for example, Greene (1905a) described 23 new species, of which only one was maintained in a later monograph (Chambers, 1955), all the rest being identified as slight morphological variants of two common Californian species. Later that same year (Greene, 1905b), he revised *Platystemon*, a western North American genus of Papaveraceae, dividing it into 52 species. In all the current floras for this region, including the recent treatment by Clark (1993) for California, this genus is considered to be monotypic.

Probably the most notorious of Greene's efforts at species splitting was his revision of *Eschscholzia* (Papaveraceae) (Greene, 1905c). This paper assigned 112 species, mostly of his own description, to the genus. Only six of the new species of *Eschscholzia* proposed by Greene are recognized by Clark (1993) for the flora of California.

Greene's taxonomic work was highly controversial at the time it was published, and many of his contemporaries were scathing in their criticism (*e.g.*, Brandegee, 1894). Willis L. Jepson, whose views on species were more conservative than Greene's, went to considerable lengths to gather experimental evidence refuting the latter's *Eschscholzia* classification. In a seldom cited study that coincided exactly with the first of Turesson's genecological publications, Jepson (1922) described "cultural and field studies" of *E. californica* which led him to merge 39 of Greene's described species into this one variable taxon. Jepson's field observations and experiments with transplanted population samples of *Eschscholzia*, extending over an 18-year period, led him to conclude that the variations given species rank by Greene were mainly the result of the morphological plasticity of many vegetative and floral traits. Flowers of strikingly different color and size might be produced on the same individual at different times in the growing season, for example. In traits that were less affected by the plants' growing conditions, Jepson noted the existence of continuous intergradation and lack of character correlations over the range of the species. A later study of ecotypic variation in *E. californica* led Cook (1962) to describe the species as composed of "a graded patchwork of distinctive populations, each nicely adapted to the particular local conditions for existence."

The conflict between Greene and his contemporaries was not along lines that divided "field naturalists" from "herbarium taxonomists" nor "descriptive botanists" from "experimentalists"; rather it occurred within the sphere of traditional morphology-based plant taxonomy. It was only in the next phase of systematics, initiated by Turesson and his followers, that taxonomy was able to penetrate the barrier of the phenotype and study the underlying genetic, ecological, and cytological causes of plant variation.

THE "EVOLUTIONARY TAXONOMY" OF HALL AND CLEMENTS

Among the individuals who actively sought to reform plant taxonomy during the period when experimental methods were beginning to come to the forefront, the American ecologist Fredric E. Clements must be given prominent mention. Clements' view that taxonomy needed to be changed "from a field overgrown with personal opinions to one in which scientific proof is supreme" was vigorously propounded in a classic publication titled *The Phylogenetic Method in Taxonomy* (Hall & Clements, 1923 [quote from p. 3]). To the authors of this work, taxonomy was to be more than descriptive botany ("which is merely a cataloguing of all known forms, with little regard to development and relationship"); it should classify plants entirely by their evolutionary, hence phylogenetic, relationships. Ecology, defined in a very broad sense, was to provide the methods for determining phylogenetic relationship (an ecologist was "anyone that employs quantitative and experimental methods in the study of plants or animals in the natural habitat"). Classification, according to Hall and Clements, was to be a synthesis of relationships based on analysis of large samples from the natural ranges of species, treated statistically, and from the results of transplant experiments. The authors were clear in their belief that by this approach most of the species segregates proposed by taxonomic splitters like Greene and Rydberg, mentioned above, would be identified as simply "variads" (that is, variations) of more inclusive natural species to be defined by commonality of descent. Sampling must include "individuals representing the whole range of variation and adaptation" to be observed and measured "in controlled natural habitats as well as in garden and greenhouse" (Hall & Clements, 1924:19).

The vigor of Clements' critique of the practices of many orthodox taxonomists probably delayed general acceptance of his views (Hagen, 1983, 1984). However, his concept of plant species as evolutionarily derived clusters of races ("variads") adapted to a variety of natural habitats clearly echoed Turesson's genecological theories and the experimental categories of ecotypes and ecospecies. Hall & Clements (1924) cited Turesson (1922b) but did not discuss his work. The weak point in Clements' own research with ecological variation was his overemphasis on environmentally induced changes, which he apparently took to be heritable in the Lamarckian sense. It fell to his successors, notably the team of Jens Clausen, David D. Keck, and William M. Hiesey, to disprove once and for all the inheritance of direct environmental modifications in plants. Although Clements had reported the experimental transformation of lowland species into alpine ones, and *vice versa*, along a transect in the Rocky Mountains, Clausen and his coworkers ultimately dismissed these claims as unreliable and contrary to the results of their own experiments and those of Kerner and Turesson (Clausen, Keck, & Hiesey, 1940; Hiesey, 1940).

TURESSON AND THE ADVENT OF GENECOLOGY

The term genecology was coined by Turesson (1923) for the study of "species ecology," as a parallel to autecology and synecology, the study of the ecology of individuals and communities, respectively. However, the term could not escape almost immediate association with the *genetic* properties of plant populations (Heywood, 1959), particularly since Turesson himself entitled one of his seminal papers "The genotypical response of the plant species to the habitat" (Turesson, 1922b). This broadened concept of the scope of genecology (Davis & Heywood, 1963, Chap. 12; Heslop-Harrison, 1964) was objected to by Langlet (1971), who insisted that studies of the inheritance of the adaptive properties of plant populations be considered a part of genetics, not the ecological field of genecology. Such disagreements are now moot in light of the integration of genecology into plant population biology as a component of "the new ecological genetics" (Antonovics, 1976).

In Turesson's early writings on the subject of genecology, we can discern some of the same desire for reform that had motivated Hall and Clements in their criticism of the taxonomic practices of the time. In a brief on behalf of his newly defined science of genecology, Turesson (1923) pointed out that "the genecological units do not necessarily — and probably quite often do not — coincide with the units of the systematists." The reason for this, he said, was that "(f)rom the point of view of traditional systematism a species is composed of a *forma genuina*, and deviations are subordinated under this type as varieties and forms of 'less systematic' value." This point of view was, to Turesson, "untenable." "There is another point of divergence in the unit conception of genecology and systematism, *viz.* the tendency of the latter to split the species into smaller ones, thus creating a swarm of units which all rank as 'species.' From a genecological point of view this is to mistake the bricks of a building for the building itself" (Turesson, 1923:173). Turesson considered "the species problem . . . to be in large measure an ecological problem" (Turesson, 1922a), and it had become his goal to "arrive at an understanding of the Linnean species from an ecological point of view" (*op. cit.*, p. 102). It was inevitable, therefore, that taxonomy and genecology would be closely linked from the beginning. Turesson's historical contribution to botany was the combining of ecology, genetics, and systematics into a coherent new field of study. With plant taxonomy thus changed from a principally descriptive science into an at least partly experimental one, it could become an important element in the middle-twentieth century "new evolutionary synthesis" (Huxley, 1940, 1942; Stebbins, 1950).

THE BEGINNINGS OF EXPERIMENTAL TAXONOMY

Experimental taxonomy and biosystematics are the two most commonly used terms for research incorporating ecological methods, along with those of genetics, cytology, and plant breeding, to study plant variation and evolution. Such research inevitably focuses at the level of species and their constituent populations. Ecological adaptation, as shown by Turesson, is characteristic of races — hence, populations — of plants; likewise, the genetic intercompatibility required for crossing experiments and cytological analysis of hybrids was generally found to be limited to closely related species and their infraspecific variants. Questions of relationship of taxa above the species level could generally not be studied by experimental methods and therefore remained the province of comparative morphology until the recent advent of molecular systematics (Crawford, 1990; Soltis, *et al.*, 1991). Furthermore, the types of plants that are amenable to cultivation under controlled conditions, and to experimental hybridizations and cytological analysis, are at best a small fraction of the flora. Thus, although biosystematics yielded significant insights on processes of microevolution and speciation, its effect in advancing the whole of plant taxonomy was necessarily quite limited.

Biosystematics, both by its methods and its theoretical underpinnings, rapidly gained influence following Turesson's publications, and the field was the subject of intense debate and review, especially in the decades between 1940 and 1970 (*e.g.*, Camp & Gilly, 1943; Constance, 1953; Davis & Heywood, 1963; Faegri, 1937; Gregor, 1944, 1963; Heslop-Harrison, 1964; Heywood, 1959, 1973; Kruckeberg, 1969a, 1969b; Mason, 1950; Ornduff, 1969; Raven, 1974; Snaydon, 1973; Stebbins, 1950; Turrill, 1946; Valentine, 1978; Valentine & Löve, 1958; Wilkins, 1968). Much of the early debate centered on the question of whether the "categories of experimental taxonomy" could readily be translated into the traditional categories of the Linnean hierarchy — genus, species, subspecies, and varietas. Although it is no longer a point of contention, having been answered in the negative by a majority of taxonomists, this question is reviewed briefly below because of its relevance to the implied expectation that experimental methods might change taxonomy from a subjective, descriptive science into a quantitative and putatively objective one.

The genecological categories of 'ecotype,' 'ecospecies,' and 'cenospecies' were coined by Turesson so as to provide an experimentally based ecological classification that was independent of the morphologically based categories of formal taxonomy (the so-called Linnean classification). The meanings of these terms changed somewhat with time (Davis & Heywood, 1963, Chap. 12) to incorporate the idea of genetic compatibility among populations, *i.e.*, their capacity (or lack of it) to "interchange their genes without detriment to the offspring" (Clausen, Keck, & Hiesey, 1940:vii). Ecotypes were thought of as the product of a sorting out of adapted genotypes from the pool of freely intermingling genes constituting an ecospecies; the cenospecies was the still more inclusive gene pool consisting of all those ecospecies that "may exchange genes among themselves to at least a limited extent" but which "are separated from one another by internal barriers that prevent such free gene exchange" (Clausen, Keck, & Hiesey, *loc. cit.*). The great attraction of this approach to studying plant variation patterns was that the criteria for classification appeared to be open to experimental verification; they were quantifiable and objective, in other words. The highly successful research program by Jens Clausen and his coworkers at the Carnegie Institution laboratories at Stanford University was built around the goal of identifying and characterizing ecotypes and ecospecies as "the natural ecologic-evolutionary units" which are "built from many genes into complex morphologic-physiologic-cytogenetic systems" (Clausen, Keck, & Hiesey, 1940:427). By analysing ecotypic differentiation (through transplant experiments) and barriers to gene exchange (through artificial hybridizations) within numerous different plant groups, they hoped "to discover principles that govern the distribution of plants and their organization into natural units" (*op. cit.*, p. 1).

Although the "experimental studies on the nature of species" by Clausen and his colleagues were a model for aspiring plant biosystematics of the time, few institutions could provide the facilities and financial support for transplant experiments in a range of differing environments. Due to practical limitations, therefore, research in experimental taxonomy mostly involved growing population samples in a greenhouse or single outdoor plot, making the necessary cross-pollinations within and between species, and raising successful hybrid progeny through the F_1, and perhaps F_2, generation. Cytological investigations on parents and hybrid offspring were essential in order to relate measurements of fertility (seed-set, pollen stainability) to the pairing and segregation of chromosomes observed at appropriate stages of meiosis. In combination with field studies of natural populations, often involving "mass collections" and techniques for detecting introgressive hybridization (Anderson, 1949), this became the research protocol for traditional plant biosystematics (Solbrig, 1970; Stace, 1980).

THE TAXONOMIC INTERPRETATION OF BIOSYSTEMATIC DATA

It is easy to see, in retrospect, why some of the early practitioners of genecology expected that the formal classification of many plant groups could readily accommodate their experimental results. After all, we do consider genera, species, and subspecies to be "natural units," the products of evolutionary processes which we can, in principal, understand and describe. The already quoted goal by Clausen and his colleagues to "discover principles that govern the . . . organization [of plants] into natural units" clearly expresses the hopes of experimental taxonomy. Transferring genecological data directly into the realm of formal taxonomy, however, has rarely proved to be practical. Despite the conscious attempt by Clausen, Keck, & Hiesey (1940) — themselves a team of geneticist, taxonomist, and physiologist — to classify their plants according to Turesson's genecological principles, the correspondence of formal taxonomy with experimentally derived population categories was approximate at best. For example, "[m]ore than one taxonomic species . . . are usually involved in a cenospecies, which sometimes corresponds to a taxonomic section or a genus"; and "[e]cospecies . . . may or may not correspond to the Linnaean taxonomic species." (*op. cit.*, p. vii). At the infraspecific level, "[o]ne or more ecotypes may be contained within one species; if these are morphologically distinguishable, we classify them as taxonomic subspecies" (*loc. cit.*).

There is a rich literature involving the debate among systematists, during the formative period of experimental taxonomy, about the relationship between biosystematics and formal taxonomy (Hagen, 1981, 1983, 1984). The main points of controversy appear to the present reviewer to be reducible to the following. Firstly, the goals of taxonomy and biosystematics are not the same. Mason (1950) expressed this well in his view that taxonomy comprises the more inclusive endeavor of "the synthesis of interrelationship", with biosystematics being placed as just one element in a division of taxonomy (called 'systematic botany') that provides the data from which relationship can be evaluated. This is expressed in another way by saying that biosystematics deals with evolutionary processes, which themselves may produce quite different taxonomic outcomes in different plant groups (Valentine & Löve, 1958; Constance, 1964). Secondly, at the infraspecific level, it was pointed out that ecological variation is more often clinal than discontinuous, and furthermore that the physiological traits most important to ecotypic adaptation are often not distinguishable morphologically (Heywood, 1959). Ecological variation occurs on many different scales and involves plant responses to ranges of different and overlapping environmental factors (Stebbins, 1950, Chap. 2; Davis & Heywood, 1963, Chap. 12). According to Heslop-Harrison (1964), attempts to assimilate genecology into taxonomy represent "an unfortunate trend" due to their difference of aims.

A third criticism of experimental taxonomy was that it overemphasized hybrid fertility and sterility as criteria for the taxonomic delimitation of plant species. Although it might have seemed reasonable at the time to try to apply the "biological species concept" (proposed and most strongly defended by zoologists) to populations of higher plants, this is no longer considered to be either practical or scientifically justified (V. Grant, 1957; Lewis, 1966; Levin, 1979; Raven, 1980). Isolating mechanisms are quite diverse in plants (Levin, 1978); the elucidation of this diversity, in fact, is one of the significant contributions of biosystematics research. In that research, genetic incompatibility and hybrid sterility, often correlated with chromosomal differences such as polyploidy, aneuploidy, and structural rearrangements (both evident and cryptic), were detected both within and between taxonomic species in many plant genera. That such barriers can easily arise and persist *within a single species* cannot be reconciled with the dogma of "biological species," yet example after example was discovered (*e.g.*, in *Clarkia* [Lewis, 1953], *Mimulus* [Vickery, 1978], *Lasthenia* [Ornduff, 1966], and many others). Even the work of Jens Clausen, who firmly espoused the idea of gradual species evolution from ecotype to ecospecies to cenospecies by slow accumulation of genetic and physiological differences, provided contrary examples of intraspecific populations isolated by rapidly evolved genetic and chromosomal barriers (*e.g.*, *Holocarpha*; J. Clausen, 1951).

Among the plant genera that were early objects of classic biosystematic studies are *Crepis* and *Achillea*, family Asteraceae. In the taxonomic treatments of these genera, quite different use was made of biosystematic categories. For *Crepis*, Babcock (1947:39) stated that due to lack of appropriate studies "ecotype, ecospecies, cenospecies, and comparium are seldom mentioned in the present work." Additionally, "if we consider hybrid sterility alone, we are compelled to recognize, as comparia, groups of extremely close and relatively remote species within the same genus" (*op. cit.*, p. 40). He therefore viewed the ability to make sterile hybrids as "another illustration of the inadequacy of a single criterion to serve as the basis for the systematic classification of organisms" (*loc. cit.*).

In *Achillea*, studies by Clausen, *et al.* (1948) revealed a textbook example of "at least eleven statistically detectable climatic races" along a 200-mile transect in central California. Because of a presumed crossing barrier between tetraploid (2n = 36) and hexaploid (2n = 54) populations, two species — *A. lanulosa* and *A. borealis*, respectively — were recognized. When it was later discovered (Hiesey & Nobs, 1970) that hybrids and segregating later-generation progeny could be obtained from crosses between tetraploids and hexaploids, all were combined within the circumpolar taxon *A. millefolium* (Nobs, 1960). Nobs' classification recognized 10 named varieties of *A. millefolium* in the Pacific States, comprising most of the earlier-detected climatic races, each with its characteristic 4x or 6x chromosome number. In his key to the varieties, Nobs used such features as plant height, leaf texture, pubescence, and degree of leaf dissection, which are quite likely to be ecotypically adaptive traits (Gurevitch, 1992). On the other hand, Tyrl (1975) showed that the two chromosome levels in *A. millefolium* are distributed in a patchwork pattern in the Pacific States and are often closely adjacent in the same habitats. He, therefore, considered the tetraploids and hexaploids to be two intermingling phases of a single highly polymorphic and genetically heterogeneous species. The attempt by Nobs to divide the continuum of variation into taxonomic varieties was not practical, in the opinion of this reviewer, because the pattern of racial differentiation is essentially continuous and because chromosome number has no taxonomic utility.

Another plant whose ecotypic relations were studied at the three transplant gardens of the Carnegie Institution laboratory of Clausen *et al.* is the grass *Deschampsia cespitosa* (Lawrence, 1945). This species parallels *Achillea millefolium* in its circumpolar range and extreme habitat differentiation. Lawrence detected five climatic ecotypes among his eight sampled populations, but he saw no correlation between these ecotypes and the three morphologically defined subspecies in his taxonomic treatment for North American *D. cespitosa.* He concluded (Lawrence, 1945:311-312) that "[t]he basic ecologic unit is therefore not the taxonomic species or subspecies but the ecotype . . . . [f]undamentally, the ecotype is not a taxonomic unit." This work proves that Clausen and his colleagues were flexible in their integration of genecological principles with practical taxonomic decisions.

It might be useful for certain special purposes if ecological variants within species could be categorized and named, and as pointed out by Davis & Heywood (1963) among others, this can be done informally by means of deme terminology (Gregor, 1963). When variation in a particular species allows ecotypes to be defined regionally (for example on the basis of adaptations to climatic zones or peculiar substrates), subspecific taxa may be recognizable which are distinct morphologically and geographically. The description and naming of such infraspecific taxa offers no taxonomic problems (Heslop-Harrison, 1963, Chap. 4). Several authors have advocated that the formal taxonomic category of variety (Latin, *varietas*) be used for the naming of Turessonian ecotypes (Heywood, 1959; Stace, 1976; Valentine, 1978; but see the counter arguments by Boivin, 1960; Merxmueller, 1963). Discussions of this question tend to become entangled with arguments over how to define the categories of subspecies and variety themselves, a perennial source of polemics among taxonomists (R. Clausen, 1941; Fosberg, 1942). It is surely possible, nonetheless, if a fine resolution classification is deemed practical and useful for a particular group, to adopt the subspecies category for morphologically well defined geographic subdivisions of a species and the varietal category for ecotypic differences on a smaller scale, taking advantage of multivariate statistics to correlate data from a wide range of attributes — morphological, physiological, and biochemical (Snaydon, 1984).

With respect to the choices that a taxonomist makes, for example whether or not to give varietal names to ecological races, it is well to keep in mind that "taxonomy can never be more than approximation in relation to the complex patterns of variation found in nature" (Heywood, 1959), and that any classification of a group of organisms should "represent their relative relationship . . . in a simplified, formalized, easily comprehended form" (Lewis, 1963). In making taxonomic decisions, it is necessary to strike a balance between the utility of a simple general-purpose classification versus the detail and precision of an elaborate, but perhaps impractical one, based on subtle cytogenetic or genecological distinctions.

THE ACHIEVEMENTS OF BIOSYSTEMATICS

Early in this review Turesson was identified as a founder of the field of biosystematics, of which experimental taxonomy was considered an approximate synonym. In general terms, the goals of this field were to elucidate the evolutionary processes that determine patterns of relationship among plant populations, and furthermore to develop classifications that best express this evolutionary relationship. In practical terms, however, two principal conceptual elements from Turesson's writings came to predominate in biosystematics research. One was the theory of ecotypic differentiation, or genecology at the level of races and local populations. Work in this area had its ultimate impact mainly in ecology, contributing strongly to the development of present day plant population biology (Raven, 1979). The second element from Turesson was the theory of ecospecies and cenospecies, or species as evolutionary units within which genes were freely recombined but between which there existed barriers to genetic exchange. It was from the latter contribution that the "physiologic-genetic species concept" of Clausen and coworkers developed (Clausen, 1951, Chap. 8; Hiesey, 1964). More importantly, it was here that cytogenetics blossomed as a taxonomic tool, profoundly enhancing the analysis and classification of "difficult" genera whose evolution involved polyploidy, chromosome rearrangements, hybridization, and peculiarities of breeding system such as autogamy and apomixis (Stebbins, 1950; V. Grant, 1981).

Briggs & Walters (1984, Chap. 12) generalized, under five categories, the "success of biosystematics." These were: (1) the testing for reproductive isolation, in all its forms; (2) the analysis and taxonomic disposition of polyploid complexes; (3) the study of interspecific hybridization and its link with polyploidy; (4) the elucidation of apomictic complexes and their need for peculiar taxonomic treatment; and (5) the recognition of habitual inbreeding as a cause of certain taxonomically critical variation patterns. The present author, from his experience with various taxonomically difficult plant genera, has every reason to "second" the evaluation by Briggs and Walters. I note that all these contributions came from the species-level theoretical input of Turesson, rather than from the ecotype level (as defined in the preceding paragraph). These successes, if such they be, are a tribute to the insight by Clements, Hall, Turesson, and other early-twentieth century systematists and ecologists, that taxonomy would be revitalized if, through observation and experiment, it could incorporate the study of evolutionary processes. In the event, this occurred by taxonomists making use not only of ecology but especially of the then nascent fields of plant genetics and cytogenetics, and to a lesser extent plant physiology. Today's plant biosystematists are inheritors of the Turesson tradition; their work continues to expand our knowledge of the processes and patterns of plant evolution and phylogeny (W. Grant, 1984; Raven, 1974, 1976, 1979).

Literature cited

Anderson, Edgar. 1949. *Introgressive Hybridization.* John Wiley & Sons, New York. 109 pp.

Antonovics, Janis. 1976 The input from population genetics: "The new ecological genetics." *Syst. Bot.* 1:233-245.

Babcock, Ernest B. 1947. The genus *Crepis.* Part I. The taxonomy, phylogeny, distribution, and evolution of *Crepis. Univ. California Publ. Bot.* 21:xii + 1-198.

Boivin, Bernard. 1960. A classical taxonomist looks at experimental taxonomy. *Rev. Canadian Biol.* 19:435-444.

Brandegee, Katharine. 1894. [Review of] Manual of the Bay Region Botany, by Edward Lee Greene. *Zoe* 4:417-420.

Briggs, D., and S. M. Walters. 1984. *Plant Variation and Evolution*, 2nd ed. Cambridge Univ. Press, Cambridge, UK. 412 pp.

Camp, Wendell H., and Charles L. Gilly. 1943. The structure and origin of species. *Brittonia* 4:323-385.

Chambers, Kenton L. 1955. A biosystematic study of the annual species of *Microseris. Contr. Dudley Herb., Stanford* 4:207-312.

Clark, Curtis. 1993. Papaveraceae. Pages 810-816 *in* James C. Hickman, ed., *The Jepson Manual: Higher Plants of California*. Univ. California Press, Berkeley.

Clausen, Jens. 1951. *Stages in the Evolution of Plant Species*. Cornell Univ. Press, Ithaca, NY. 206 pp.

Clausen, Jens, David D. Keck, and William M. Hiesey. 1940. Experimental studies on the nature of species. I. Effect of varied environments on Western North American plants. *Carnegie Inst. Washington, Publ.* 520. 452 pp.

Clausen, Jens, David D. Keck, and William M. Hiesey. 1948. Experimental studies on the nature of species. III. Environmental responses of the climatic races of *Achillea. Carnegie Inst. Washington, Publ.* 581. 129 pp.

Clausen, Robert T. 1941. On the use of the terms "subspecies" and "variety." *Rhodora* 43:157-167.

Constance, Lincoln. 1953. The role of plant ecology in biosystematics. *Ecology* 34:642-649.

Constance, Lincoln. 1964. Systematic botany — an unending synthesis. *Taxon* 13:257-273.

Cook, Stanton A. 1962. Genetic system, variation, and adaptation in *Eschscholzia californica. Evolution* 16:278-299.

Crawford, Daniel J. 1990. *Plant Molecular Systematics. Macromolecular Approaches*. John Wiley & Sons, New York. 388 pp.

Davis, P. H., and V. H. Heywood. 1963. *Principles of Angiosperm Taxonomy*. Oliver & Boyd, London, UK. 556 pp.

Faegri, Knut. 1937. Some fundamental problems of taxonomy and phylogenetics. *Bot. Rev.* (Lancaster) 3:400-423, 451-456.

Fosberg, F. Raymond. 1942. Subspecies and variety. *Rhodora* 44:153-167.

Grant, Verne. 1957. The plant species in theory and practice. Pages 39-80 *in* Ernst Mayr, ed., *The Species Problem*. American Association for the Advancement of Science, Washington, DC.

Grant, Verne. 1981. *Plant Speciation,* 2nd ed. Columbia Univ. Press, New York. 563 pp.

Grant, William F. (ed.). 1984. *Plant Biosystematics*. Academic Press, Toronto. 674 pp.
Greene, Edward L. 1905a. A new study of *Microseris*. *Pittonia* 5:4-16.
Greene, Edward L. 1905b. *Platystemon* and its allies. *Pittonia* 5:139-194.
Greene, Edward L. 1905c. Revision of *Eschscholtzia*. *Pittonia* 5:205-292.
Gregor, J. W. 1944. The ecotype. *Cambridge Biol. Rev.* 19:20-30.
Gregor, J. W. 1963. Genecological (biosystematic) classification: The case for special categories. *Regnum Vegetabile* 27:24-26.
Gurevitch, Jessica. 1992. Sources of variation in leaf shape among two populations of *Achillea lanulosa*. *Genetics* 130:385-394.
Hagen, Joel B. 1981. Experimental taxonomy, 1930-1950: The impact of cytology, ecology, and genetics on ideas of biological classification. *Ph.D. Thesis*, Oregon State University, Corvallis. 226 pp.
Hagen, Joel B.. 1983. The development of experimental methods in plant taxonomy, 1920-1950. *Taxon* 32:406-416.
Hagen, Joel B. 1984. Experimentalists and naturalists in twentieth-century botany: Experimental taxonomy, 1920-1950. *Jour Hist. Biol.* 17:249-270.
Hall, Harvey M., and Fredric E. Clements. 1923. The Phylogenetic Method in Taxonomy. The North American Species of *Artemisia, Chrysothamnus*, and *Atriplex*. *Carnegie Inst. Washington, Publ.* 326. 355 pp.
Heslop-Harrison, J. 1963. *New Concepts in Flowering Plant Taxonomy*. Heinemann, London. 134 pp.
Heslop-Harrison, J. 1964. Forty years of genecology. Pages 159-264 *in Advances in Ecological Research*, Vol. II.
Heywood, V. H. 1959. The taxonomic treatment of ecotypic variation. Pages 87-112 *in Function and Taxonomic Importance*. The Systematics Association, Publ. No. 3. Academic Press, London, UK.
Heywood, V. H.. 1973. Ecological data in practical taxonomy. Pages 329-347 *in Taxonomy and Ecology*. The Systematics Association, Special Vol. No. 5. Academic Press, London, UK.
Hiesey, William M. 1940. Environmental influence and transplant experiments. *Bot. Rev.* (Lancaster) 6:181-203.

Hiesey, William M. 1964. The genetic-physiologic structure of species complexes in relation to environment. Pages 437-445 *in Genetics Today*. Proc. XI Internat. Congr. Genetics. Pergamon Press, London.

Hiesey, William M., and Malcolm A. Nobs. 1970. Genetic and transplant studies on contrasting species and ecological races of the *Achillea millefolium* complex. *Bot. Gaz.* 131:245-259.

Huxley, Julian S. (ed.) 1940 *The New Systematics*. Clarendon Press, Oxford. 334 pp.

Huxley, Julian S. 1942. *Evolution: The Modern Synthesis*. Harper and Sons, New York. 645 pp.

Jepson, Willis L. 1922. *Eschscholtzia*. Pages 564-575 *in A Flora of California*. Vol. I, pt. 7. Univ. California Press, Berkeley.

Kruckeberg, Arthur R. 1969a. The implications of ecology for plant systematics. *Taxon* 18:92-120.

Kruckeberg, Arthur R. 1969b. Ecological aspects of the systematics of plants. Pages 161-203 *in Systematic Biology*. Nat'l Acad. Sci., Washington, D.C. Publ. No. 1692.

Langlet, Olof. 1971. Two hundred years [of] genecology. *Taxon* 20:653-722.

Lawrence, William E. 1945. Some ecotypic relations of *Deschampsia caespitosa*. *American Jour. Bot.* 32:298-314.

Levin, Donald A. 1978. The origin of isolating mechanisms in flowering plants. Pages 185-317 *in* Max K. Hecht, William C. Steere, & Bruce Wallace, eds., *Evolutionary Biology*, Vol. 11. Plenum Press, New York.

Levin, Donald A. 1979. The nature of plant species. *Science* 204: 381-384.

Lewis, Harlan. 1953. The mechanism of evolution in the genus *Clarkia*. *Evolution* 7:1-20.

Lewis, Harlan. 1963. The taxonomic problem of inbreeders. *Regnum Vegetabile* 27:37-44

Lewis, Harlan. 1966. Speciation in flowering plants. *Science* 152: 167-172.

Mason, Herbert L. 1950. Taxonomy, systematic botany and biosystematics. *Madroño* 10:161-192.

Mayr, Ernst. 1987. The ontological status of species: Scientific progress and philosophical terminology. *Biol. and Philos.* 2:145-166.

Merxmueller, Hermann. 1963. The incompatibility between formal taxonomic recognition of units and their biosystematic definition. *Regnum Vegetabile* 27:57-62.

Nobs, Malcolm A. 1960. *Achillea.* Pages 390-391 *in* Roxana S. Ferris, ed., *Illustrated Flora of the Pacific States*, Vol. IV. Stanford Univ. Press, Stanford, CA.

Ornduff, Robert. 1966. A biosystematic survey of the goldfield genus *Lasthenia. Univ. California Publ. Bot.* 40:1-92.

Ornduff, Robert. 1969. The systematics of populations in plants. Pages 104-128 *in*: *Systematic Biology*. Nat'l Acad. Sci., Washington, D.C. Publ. No. 1692.

Raven, Peter H. 1974. Plant systematics, 1947-1972. *Ann. Missouri Bot. Gard.* 61:166-178.

Raven, Peter H. 1976. Systematics and plant population biology. *Syst. Bot.* 1:284-316.

Raven, Peter H. 1979. Future directions in plant population biology. Pages 461-481 *in* Otto T. Solbrig, Subodh Jain, George B. Johnson & Peter H. Raven, eds., *Topics in Plant Population Biology.* Columbia Univ. Press, New York.

Raven, Peter H. 1980. Hybridization and the nature of species in higher plants. *Canadian Bot. Assn. Bull.* 13(1)(suppl.):3-10.

Rydberg, Per Axel. 1929. Fabaceae, Subtribe Astragalanae. Pages 251-462 *in North American Flora*, Vol. 24, Pts. 5-7. New York Botanical Garden, NY.

Snaydon, R. W. 1973. Ecological factors, genetic variation and speciation in plants. Pages 1-29 *in Taxonomy and Ecology.* The Systematics Association, Special Vol. No. 5. Academic Press, London.

Snaydon, R. W. 1984. Infraspecific variation and its taxonomic implications. Pages 203-218 *in* V. H. Heywood & D. M. Moore, eds., *Current Concepts in Plant Taxonomy.* The Systematics Association Special Vol. No. 25. Academic Press, London.

Solbrig, Otto T. 1970. *Principles and Methods of Plant Biosystematics.* Macmillan Co., Toronto. 226 pp.

Soltis, Pamela S., Douglas E. Soltis, and Jeff J. Doyle. 1991. *Plant Molecular Systematics.* Chapman and Hall, New York. 448 pp.

Stace, Clive A. 1976. The study of infraspecific variation. *Curr. Adv. Plant Sci.* 8:Commentary 23.

Stace, Clive A. 1980. *Plant Taxonomy and Biosystematics*. Edward Arnold, London, UK. 279 pp.

Stebbins, G. Ledyard, Jr. 1950. *Variation and Evolution in Plants*. Columbia Univ. Press, New York. 643 pp.

Turesson, Göte. 1922a. The species and the variety as ecological units. *Hereditas* 3:100-113.

Turesson, Göte. 1922b. The genotypical response of the plant species to the habitat. *Hereditas* 3:213-350.

Turesson, Göte. 1923. The scope and import of genecology. *Hereditas* 4:171-176.

Turrill, W. B. 1946. The ecotype concept: A consideration with appreciation and criticism, especially of recent trends. *New Phytol.* 45:34-43.

Tyrl, Ronald J. 1975. Origin and distribution of polyploid *Achillea* (Compositae) in Western North America. *Brittonia* 27:187-196.

Valentine, D. H. 1978. Ecological criteria in plant taxonomy. Pages 1-18 *in* H. E. Street, ed., *Essays in Plant Taxonomy*. Academic Press, London., UK.

Valentine, D. H. and Askell Löve. 1958. Taxonomic and biosystematic categories. *Brittonia* 10:153-166.

Vickery, Robert K., Jr. 1978. Case studies in the evolution of species complexes in *Mimulus*. Pages 405-507 *in* Max Hecht, William C. Steere, & Bruce Wallace, eds., *Evolutionary Biology*, Vol. 11. Plenum Press, New York.

Wilkins, D. A. 1968. The scale of genecological differentiation. Pages 227-239 *in* V. H. Heywood, ed., *Modern Methods in Plant Taxonomy*. Academic Press, London, UK.

Williams, Roger L. 1984. *Aven Nelson of Wyoming*. Colorado Assoc. Univ. Press, Boulder. 407 pp.

Ecotypic Variation in Response to Serpentine Soils[1]

Arthur R. Kruckeberg
Department of Botany, AJ-30
University of Washington, Seattle, WA 98195

Turesson's conception of the ecotype is elegant in its simplicity: The ecotype is the genotypical response of a species (or a population) to the habitat (Turesson, 1922 a, b). And thus Turesson's genecology is the study of the ecotypic variation within a species. Turesson's examples of ecotypes were those responding to differences in climate, though Turesson later recognized edaphic ecotypes as well. His ecotype concept has undergone a metamorphosis from the landmark papers of 1922 to the present. The first modification, suggested by Stebbins (1950) and Heslop-Harrison (1953) was to employ the term and concept, ecotypic variation, in preference to the more typological (*sensu* Mayr, 1963) notion of the ecotype. Stebbins reasoned that discrete units, ecotypes, are less likely to be recognized in nature than a graded, more or less continuous genotypic response to varying environments. Support for Stebbin's shift to the term, ecotypic variation, came soon after his published argument. In 1951, Kruckeberg found that within the California Sierran foothill "ecotype" of *Achillea millefolium,* further ecotypic differentiation could occur. The substrate diversity in the foothill range of yarrow was manifestly great. It included serpentine and non-serpentine soils, to which yarrow showed contrasting tolerance. Could this be construed as "ecotypes within ecotypes"? Or better to view it as ecotypic variation in response to both climate and substrate?

Further evolution of the ecotype concept has occurred. The notion of local race is only a different way of acknowledging ecotypic variation within a species. The same can be said for the forester's well-recognized provenance type; it is simply "race" (or ecotype) in French trappings.

Nowadays studies of infraspecific variation marches under the

[1] I dedicate this paper to the three pioneers of genecology in California: Jens Clausen, David Keck, and William Hiesey (1940, 1948), whose pioneer work guided me down the genecological path.

banner of population biology. Whether or not population biologists formally recognize the ecotype, ecotypic variation, race or provenance, is probably irrelevant. What is relevant to the theme of this symposium is that the population biologist is carrying on the Turesson tradition, but using newer techniques. Further, contemporary research is more preoccupied with the genetic and evolutionary consequences of genotypic response to habitat.

The present paper reviews a chapter in genecology that had its inception in the 1950s. Largely through the stimulus of two milestone papers by Herbert Mason (1946a,b), on the edaphic factor in narrow endemism, botanists in California began to look at the plant responses to serpentine soils (Kruckeberg, 1951, 1954, and review of 1980).

Studies on serpentine soils in California and in Washington State focused on the issue of variation in tolerance to serpentine within species (Kruckeberg, *ibid.*). The present review will reexamine these earlier studies, will add newer research, and will look at the highly similar plant responses to heavy metals.

SERPENTINE RACES — THE ECOTYPIC RESPONSE TO THE EDAPHIC FACTOR

In any region where there are sharp (or even subtle) contrasts in substrates, floras will respond in a variety of ways. Landscape mosaics, edaphically induced, can display species preference, contrasts in community composition, edaphically restricted endemic species and racial differentiation within species on the contrasting soils. Serpentine vegetation strikingly illustrates this range of responses in many parts of the world (Brooks, 1987; Roberts and Proctor, 1992). I address here the ways in which species of wide distribution respond to contrasting serpentine (S) and non-serpentine (NS) soils. Some species may avoid serpentines even though they may occur widely on a variety of other substrates nearby. Other species may venture onto serpentine, as well as occurring on other adjacent substrates; in so doing two possible eco-physiological and genetic responses are possible: (1) Both S and NS populations are equally tolerant of serpentine; *i.e.*, the species shows no ecotypic differentiation. This has been demonstrated for a few western North American taxa (Digger pine, *Pinus sabiniana* [Griffin, 1965], and *Sitanion jubatum* [Kruckeberg, 1950]). (2) The NS populations are intolerant of serpentine; ecotypic differentiation has evolved in such species with serpentine-tolerant populations. Examples are from a variety of "bodenvag" or "ubiquist" species[2] from California and Washington State which have been subject to progeny testing as clones or as seedlings. Certain annuals (*Gilia capitata, Salvia columbariae, Streptanthus glandulosus*), herbaceous perennials (*Achillea millefolium, Prunella vulgaris, Fragaria virginiana*), and woody species (*Pinus* spp.) show this racial response (Kruckeberg ,1980, and Miller & Jenkinson, abstract in this symposium).

Many introduced species (mostly weeds) have wide distributions in California and the Pacific Northwest. It is to be expected that some alien taxa respond ecotypically to S and NS soils. Annual grasses are prime candidates (*Avena* spp., *Bromus* spp., *Vulpia* spp.) as are introduced annual dicotyledous (dicots) like *Erodium, Silene gallica, Hypochaeris radicata,* etc. Should edaphic races be found among these weeds, it would indicate a rapid evolutionary accommodation to serpentine, following introduction from Europe in the 18th and 19th centuries. It is also possible that the aliens were preadapted in their homelands and fortuitously found new habitats to which they were already tolerant. Still another genetic recourse for aliens is that their tolerance to S and NS soils involves no genotypic differentiation into S and NS races. Such a possibility can be looked upon as the "general purpose genotype" response (Baker, 1965). It implies that one and the same genotype can tolerate a range of edaphic environments, including serpentine. So far as I am aware there are no experimental verifications of the occurrence of serpentine races in weed species.

Besides the western North American case histories just cited, there are a number of studies on ecotypic response to serpentine from other parts of the world. In Europe, serpentine races have been detected in *Thlaspi goesingense* (Reeves & Baker, 1984), *Silene vulgaris* ssp. *montana* (Schoop-Brockmann & Egger, 1980), and in *Agrostis stolonifera* (Marrs & Proctor, 1976). In the *Agrostis* studies, the authors found that the serpentine race (two populations) had a higher Mg requirement and a greater tolerance of excess Mg; this preferential uptake of Mg confirms the earlier finding of Madhok and Walker (1969) for *Helianthus bolanderi* ssp. *exilis*. The study with *Thlaspi* focussed on the effects of Ni; both the S and NS populations (the latter from calcareous soil) were tolerant of Ni. The authors suggest

[2] Bodenvag — species that occur on and off serpentine; ubiquist species have the same broad tolerance.

that the species has a "non-specific detoxification system" for Ni. Lee, *et al.* (1983) tested 15 species occurring on New Zealand serpentines. Only the herbaceous species (7 taxa) showed ecotypic differentiation; none of the three woody species tested gave evidence of edaphic ecotypes. Three of the five naturalized grass species exhibited ecotypic differentiation. This, the authors comment, is not unexpected, since grass species are well-known in having races tolerant to stressful substances (*e.g.*, heavy metals in mine tailing substrates).

ECOTYPIC RESPONSE TO HEAVY METAL SUBSTRATES

A major component of the "serpentine syndrome" (Jenny, 1980) is the occurrence of high levels of Ni. Indeed serpentine is one of a family of metalliferous soil (Brooks, 1987; Shaw 1989). Just as with the serpentine, we expect racial differentiation in species that occur on and off other metalliferous substrates. Response to soils high in arsenic, copper, lead, and zinc should, like the Ni-containing serpentine, yield ecotypic variants. In fact, the many case studies, mostly in Europe, point to these metalliferous substrates as prime examples of rapid evolutionary accommodation — genotypic response to the highly selectional impetus of heavy metals. Metal-tolerant races have been known for many years beginning with the pioneering studies of Bradshaw and his collaborators (see reviews of Antonovics and Bradshaw, 1971, and Shaw, 1989).

Heavy metal tolerance is not only a clearcut example of ecotypic response to a severe edaphic stress. The phenomenon serves elegantly as a model evolutionary system. All those who have done research on metallophytes are led to see in this edaphic paradigm, case histories in natural selection, genetic drift, reproductive isolation and frequently the capstone event of speciation leading to edaphic endemism (see reviews by Shaw, 1989, and Kruckeberg, 1986). Such a provocative biological and geochemical system continues to attract investigation. I will draw upon some of the recent studies to illustrate this genecological model par excellence.

Two recent reviews of metal tolerance (Baker, 1987; Baker & Walker, 1990) focus on the ecophysiological and evolutionary aspects of the phenomenon. It is now well established that tolerant plants cope with heavy metals, including nickel of serpentine soils, in at least two ways: metal exclusion and metal accumulation (Baker

& Walker, 1990). With either stratagem, the evolved tolerance is derived from nearby non-tolerant populations, especially demonstrable where the metalliferous habitat is of recent origin (mine spoils and similar anthropogenic sources). Thus either type of tolerance hews to the genecological model — ecotypic differentiation within a species. Only when the metalliferous habitat has been in existence for longer stretches of geologic time, has the outcome of tolerance often proceeded to speciation — the origin of metallophyte endemics (*e.g.*, Brooks & Malaise, 1985).

Most studies in the Northern Hemisphere, mainly in Europe, have dealt with tolerance to substrates contaminated by human activity (mines, their tailings, and kindred disturbed habitats containing high levels of metals). Since most such sites are of relatively recent origin, it is especially remarkable that metal tolerance can evolve so rapidly. One to a few generations of selection of preadapted genotypes can yield tolerant races (Baker, 1987).

Should ecotypic response yielding tolerant races be the only pathway to metal tolerance? Baker (1987) reviews the exceptional cases where a species can occupy metalliferous soils without the evolution of tolerant races. He suggests that in such examples as *Typha latifolia* and *Thlaspi goesingense*, "the plants may possess a constitutional tolerance to heavy metals"; this absence of racial differentiation is suggestive of the "general purpose genotype," (term coined by H.G. Baker, 1965) to account for certain weedy, colonizing species broadly tolerant of a variety of pioneer habitats. I am not aware of cases in classical genecology dealing with climatic races of ecotypes, where the possession of broad tolerance without ecotypic variants was ever recognized, either by Turesson (1922 a, b) or by Clausen, Keck & Hiesey (1946, 1948).

Metal tolerance to contaminated soils has been detected in California, mostly on copper mine sites (Macnair, 1989; Wu & Kruckeberg, 1985). Macnair's study of tolerant forms of *Mimulus* spp. reveals that copper tolerance can evolve beyond ecotypic differentiation into a copper-tolerant endemic species. Two kinds of *Mimulus* occur on these copper mine sites in the foothills of the Sierra Nevada. *Mimulus guttatus* in its annual form coexists with a newly described derivative species, also annual *M. cupriphilus* (Macnair, 1989 for the description of the new species). Macnair, *et al.* (1989) contend that (1) *M. cupriphilus* is of recent origin (post-copper mine times), (2) that the new species is reproductively isolated from its presumed progenitor, *M. guttatus,* and (3) because of its inherent autogamy and higher seed output, the new species is better adapted to the copper mine sites that is *M. guttatus.*

In the context of genecology (genotypic response to habitat), this *Mimulus* case-history clearly indicates the possible fate of an edaphic race. While some such races might exist forever simply as ecotypic variants, we may witness the outcome of a race becoming a species — ecologically, morphologically and reproductively distinct from the immediate ancestor. Figure 1 portrays the possible fates of genetic accommodation to a unique edaphic environment (Kruckeberg,

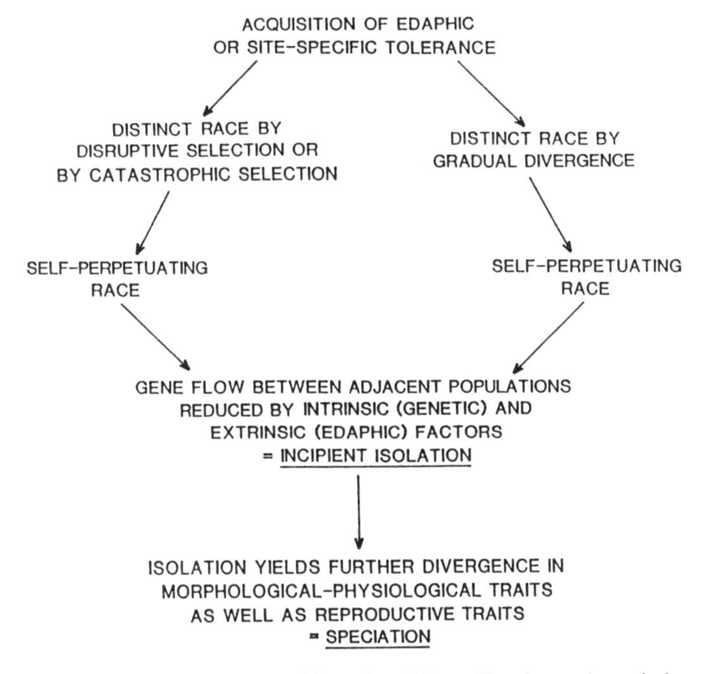

FIGURE. 1. Scheme to portray possible paths of diversification and speciation under geoedaphic (serpentine and other heavy metal) influences. From Kruckeberg, 1986.

1986). I have described five stages in the emergence of a new species, using serpentine soils as the challenging environment; the *M. cupriphilus* case appears to parallel the following evolutionary pathway.

> *Stage 0.* Some preadaptation for serpentine (or metal) tolerance exists in certain . . . populations. Stage 1. Disruptive selection effectively separates a species into serpentine-tolerant and -intolerant gene pools. *Stage 2.* Further

> genetic divergence in structural and functional traits occurs within the serpentine-tolerant part of the effectively discontinuous populations. *Stage 3.* Isolation between serpentine and non-serpentine segments of the species becomes genetically fixed; the two populations are unable to exchange genes. *Stage 4.* Further reinforcement of genetic-ecologic isolation and consequent further divergence of the serpentine population may occur, put in motion by the initial genetic discontinuity. (Kruckeberg, 1986:459).

In this model, ecotypic differentiation must be an essential early step. Examples like *M. cupriphilus* neatly illustrate rapid speciation following ecotypic differentiation; the heavy metal stimulus of mine soils can elicit the gamut of plant response - from early preadapted genotypes and edaphic races to full speciation. It is reasonable to conjecture that serpentine soil with its syndrome of stressful attributes (or any other edaphic stress) could exhibit the same gamut of response — from ecotypic variance to endemic species. In California, the serpentinicolous genera, *Streptanthus* (subgenus *Euclisia*) and *Linum* (subgenus *Hesperolinon*) are good fits to the model (sequence) (Kruckeberg, 1980, 1986; Kruckeberg & Morrison, 1983).

UNSOLVED PROBLEMS

What remains to be done? Are there still unsolved problems associated with genotypic accommodation to serpentine and other metalliferous soils? It is probably safe to assume that further demonstration of the existence of edaphic races is not of the first priority. Other case histories would show either ecotypic response or the alternative — a "general purpose" genotype with no ecotypic differentiation. But from scanning the current literature on the evolutionary and ecophysiological aspects of tolerance to serpentines and other edaphically stressful habitats, many questions still beg for answers. The most crucial is the question of the physiological and molecular basis of the tolerance. Further, we need a better understanding of the genetic basis of the tolerance — is it oligogenic or polygenic? And how is the idiosyncratic response genetically/molecularly controlled: a race (or species) either excludes the heavy metal or accumulates it (up to the remarkable level of a hyperaccumulator, with over 1000 ppm in the tissue)? I predict that this model evolutionary paradigm will continue to provoke fruitful research.

CONSERVATION OF EDAPHIC SPECIALISTS

A serpentine-tolerant race can be just as endangered as a serpentine endemic species. Witness the endangerment caused by world-wide disturbance of serpentine habitats as the result of mining activities and other human intrusions; even recreational all-terrain vehicles make impacts on serpentine barrens (Kruckeberg, 1980). Some years ago John Harper (1981) reminded us that ecological (genotypic) variants not recognized by the taxonomist merit preservation. He used the metal-tolerant race of *Agrostis tenuis* on Welsh mine tips as his example; the same case can be made for serpentine races. Not only is their preservation warranted for scientific reasons, as model evolutionary systems; metal tolerant races contribute, in a practical, land management context, to the ecological and physical stability of metalliferous soils. So while serpentine and other metallophyte endemic species are the most obvious objects of conservation, we must not overlook the protection of edaphic races.

Literature Cited

Antonovics, J., A. D. Bradshaw, and R. G. Turner. 1971. Heavy metal tolerance in plants. *Adv. Ecol. Res.* 7:1-85.

Baker, A. J. M. 1987. Metal Tolerance. *New Phytol.* 106(suppl.):93-111.

Baker, A. J. M. and P. L. Walker. 1990. Ecophysiology of metal uptake by tolerant plants. Pp. 166-177 *in* A. J. Shaw, ed., *Heavy Metal Tolerance in Plants: Evolutionary Aspects.* CRC Press, Boca Raton, FL.

Baker, H. G. 1965. Characteristics and modes of origin of weeds. Pp. 147-172 *in* H. H. Baker & G. L. Stebbins, eds., *The Genetics of Colonizing Species.* Academic Press, New York.

Brooks, R. R. 1987. *Serpentine and Its Vegetation: A Multidisciplinary Approach.* Dioscorides Press, Portland, OR.

Brooks, R. R. and F. Malaisse. 1985. *The Heavy Metal-tolerant Flora of Southcentral Africa: A Multidisciplinary Approach.* A.A. Balkema, Rotterdam/Boston.

Clausen, J., D. D. Keck, and W. M. Hiesey. 1940. Experimental studies on the nature of species. I. Effect of varied environments on western North American plants. *Carnegie Inst. Washington, Publ.* 520. 452 pp.

Clausen, J., D. D. Keck, and W. M. Hiesey. 1948. Experimental studies on the nature of species. III. Environmental responses of climatic races of *Achillea*. *Carnegie Inst. Washington, Publ.* 581. 129 pp.

Griffin, J. R. 1965. Digger pine seedling response to serpentine and non-serpentine soil. *Ecology* 46:801-807.

Harper, J. L. 1981. The meanings of rarity. Pp. 189-203 *in* H. Synge, ed., *The Biological Aspects of Rare Plant Conservation.* John Wiley & Sons, New York.

Heslop-Harrison, J. 1953. *New Concepts in Flowering-Plant Taxonomy.* W. Heinemann Ltd., London, UK.

Jenny, H. 1980. *The Soil Resource: Origin and Behavior.* Springer-Verlag, Berlin and New York.

Kruckeberg, A. R. 1950. An experimental inquiry into the nature of endemism on serpentine soils. *Ph.D. Thesis.* Univ. California, Berkeley.

Kruckeberg, A. R. 1951. Intraspecific variability in the response of certain native plants to serpentine soil. *American Jour. Bot.* 38:408-419.

Kruckeberg, A. R. 1954. The ecology of serpentine soils. III. Plant species in relation to serpentine soils. *Ecology* 35:267- 274.

Kruckeberg, A. R. 1980. California Serpentines: Flora, Vegetation, Geology, Soils, and Management Problems. *Univ. California Publ. Bot.* 78:1-180. Univ. California Press, Berkeley.

Kruckeberg, A. R. 1986. An essay: The stimulus of unusual geologies for plant speciation. *Systematic Bot.* 11:455-463.

Kruckeberg, A. R. and J. L. Morrison. 1983. New *Streptanthus* taxa (Cruciferae) from California. *Madroño* 30:230-244.

Lee, W. G., A. F. Mark, J. B. Wilson. 1983. Ecotypic differentiation in the ultramafic flora of the South Island, New Zealand. *New Zealand Jour. Bot.* 21:141-156.

Macnair, M. R. 1989. A new species of *Mimulus* endemic to copper mines in California. *Bot. Jour. Linnaean Soc.* 100:1-14.

Macnair, M. R. and Q. J. Cumbes. 1989. The genetic architecture of interspecific variation in *Mimulus. Genetics* 122:211-222.

Madhok, O. P. and R. B. Walker. 1969. Magnesium nutrition of two species of sunflower. *Plant Physiol.* 44:1016-1022.

Mason, H. L. 1946a. The edaphic factor in narrow endemism. I. The nature of environmental influences. *Madroño* 8:209-226

Mason, H. L. 1946b. The edaphic factor in narrow endemism. II. The geographic occurrence of plants of highly restricted patterns of distribution. *Madroño* 8:241-257.
Marrs, R. H. and J. Proctor. 1976. The response of serpentine and non-serpentine *Agrostis stolonifera* L. to magnesium and calcium. *Jour. Ecol.* 64:953-964.
Mayr, E. 1963. *Animal Species and Evolution.* Belknap Press, Cambridge, Mass.
Reeves, R. D. and A. J. M. Baker. 1984. Studies on metal uptake by plants from serpentine and non-serpentine populations of *Thlaspi goesingense* Halacsy (Cruciferae). *New Phytol.* 98:191-204.
Roberts, B. A. and J. Proctor (eds.). 1992. *The Ecology of Areas with Serpentinized Rocks: A World View.* Kluwer Academic Publ. Dordrecht, Netherlands.
Schoop-Brockman, I. and B. Egger. 1980. Oekologische Differenzierung bei *Silene vulgaris* L. auf saurem Silikat, Karbonat und Serpentin in der alpinen Stufe bei Davos. *Ber. Geobot. Inst. Eid. Techn. Hochschule Stiftung Rubel.* 47:50-74.
Shaw, A. J. (ed.) 1989. *Heavy Metal Tolerance in Plants: Evolutionary Aspects.* CRC Press, Boca Raton, Florida.
Stebbins, G. L. Jr. 1950. *Variation and Evolution in Plants.* Columbia Univ. Press, New York.
Turesson, G. 1922a. The species and the variety as ecological units. *Hereditas* 3:100-113.
Turesson, G. 1922b. The genotypical response of the plant species to the habitat. *Hereditas* 3:211-350.
Wu, L. and A. L. Kruckeberg. 1985. Copper tolerance in two legume species from a copper mine habitat. *New Phytol.* 99:563-570.

From the Mountains to the Prairies to the Ocean White With Foam: *Papilio zelicaon* Makes Itself at Home

Arthur M. Shapiro
Section of Evolution and Ecology, University of California, Davis, CA 95616

Papilio zelicaon, the anise swallowtail, has an extraordinary range in California — from sea level to the Sierra Nevada alpine zone, from the coastal fog belt to high desert. Its phenology varies immensely in tandem with that of its hosts and is mostly under genetic control through selection of thresholds for diapause induction and suppression. Univoltine races may occur in close proximity to multivoltine ones, and rapid evolution in seasonality may occur upon contact with introduced hosts. Citrus-associated ecotypes have apparently arisen in this century in both northern and southern California. The ecology of the northern citrus ecotype provides clues to the process of ecotype formation. Electrophoretic studies indicate that little genomic reorganization has occurred. Rather, strong selection seems to act routinely on genes affecting diapause, tailoring population to host plant phenology. Epigamic behavior limits the differentiation of ecotypes in mountainous terrain and may contribute to gene flow arising from the spread of multivoltine ecotypes and their hosts into hitherto univoltine areas. Such gene flow has the potential to give rise to disruptive selection and may be self-limiting.

"Ecotypic variation is basically a botanical concept," I wrote in 1974 even while arguing for its applicability to butterflies. Dobzhansky (1970) had noted that "there seems to be no real difference between the ecotype and the polytopic subspecies of the taxonomist." He was only partly right here. Taxonomic subspecies (polytopic or otherwise) traditionally must be recognizable as specimens, *i.e.*, phenotypically. Ecotypes are differentiated in ecological ways which have a genetic basis but which may or may not be apparent in preserved specimens. Not all phenotypic segregates of a taxonomic species need be ecotypes, and not all ecotypes need be morphologically recognizable.

The anise swallowtail, *Papilio zelicaon* Lucas, has a vast range in western North America — from British Columbia to Alberta, south throughout the Rockies and northern Great Basin, and encompassing most of California except the hottest deserts. Phenotypic variation over this vast range is absolutely minimal. The only truly distinctive geographic-phenotypic entity within *zelicaon* is the large, richly colored summer form found in lowland California. At the same time, the amount of ecological and physiological differentiation among California *P. zelicaon* populations is tremendous, even given the topographic, climatic and vegetational complexity of that state. *Papilio zelicaon* shares some features in common with the other much-studied ecotypically-differentiated western butterfly, *Euphydryas editha* Boisduval (Singer, this volume), but incorporates in its ecotypes a degree of life-history flexibility unavailable to that species.

Papilio zelicaon is a member of the "*Papilio machaon* Linnaeus group," a circumpolar (Holarctic) assemblage which appears to be a hotbed of evolutionary activity. (Sperling (1987) showed this very well for western Canada. The taxonomic literature of the group is immense and bears witness to the perplexing array of "forms," "races," "hybrids," etc. recognized phenotypically within most of the claimed morphospecies. The phenotypic stability of *zelicaon* over its range is in itself remarkable for this group!) Most of this evolutionary activity is at or below the subspecies level and appears consequent to range and ecological adjustments since deglaciation.

These tantalizing taxonomic problems alone would justify research on the *machaon* group, but the ease of handling of the animals, including particularly their ability to be hand-paired (thus circumventing the need to create conditions favorable for spontaneous matings, and allowing experimental crosses to be made *ad lib*), has contributed to their popularity as a laboratory system. So has their involvement in a classic tale of chemically-mediated host selection. The entire group is closely tied to the secondary chemistry of the Umbelliferae (most tribes) and Rutaceae. The fact that these families share their chemistry and are used more or less interchangeably, while the sister-group of the Umbels, Araliaceae, does not and is not, led to this case becoming one of the first to be elucidated by experiment rather than mere correlation (Dethier, 1941). Further advances in understanding the Umbel-*Papilio* system were carried out by Paul Feeny and his group, especially Berenbaum (*e.g.*, 1983), using the widespread eastern North American species *P. polyxenes* Fabr.

Members of the *machaon* group tend to specialize on a particular host plant locally, but to use a number of others occasionally as available. The local "favorite" is usually the most abundant potential host plant and is often, but not always, that plant which best sustains growth and development — but because there are multiple scales for assessing this, glib generalizations are not easily drawn. Detailed and careful investigations of this pattern of host utilization have been carried out by Christer Wiklund (1981 and other papers) on Fenno-scandian *machaon* and by John Thompson on *zelicaon* and *P. oregonius* W. H. Edwards in the Pacific Northwest (Thompson, 1988a, b, c). Both *machaon* and Northwestern *zelicaon* show ecotypic differentiation in host preference.

Montane populations of *P. zelicaon* in Washington feed primarily on native perennial umbellifers, including *Lomatium grayi* Coult. & Rose and *Cymopterus terebinthinus* (Hook.) T.& G. (the latter is the commonest host of alpine and subalpine Californian populations as well). In coastal Washington they use *Angelica lucida* L. Thompson (1988a) carried out oviposition preference tests which demonstrated high individual variability among females in host preference within as well as between populations. Thompson infers the existence of a host plant preference hierarchy, observing (1988a): "there is a consistent overall ranking of plant species among strains and . . . variation for host choice in these populations is expressed as variation in the degree to which females lay some eggs on lower-ranking plant species." He sums up the situation thus (Thompson, 1988b): "The *Papilio machaon* group seems to be adept at responding to the local availability of host-plant species, with most species exhibiting geographic variation in the plant species that are used as hosts (Wiklund, 1981; Thompson, 1986, 1988c). Since there is genetic variation in patterns of host selection within *Papilio* populations . . . these species appear to have the potential for local adaptation through slight shifts in the preference hierarchy . . . This does not mean, however, that all populations using different host plants necessarily differ genetically in their preferences. Since preference is . . . determined as a hierarchy . . . , ovipositing females will continue down the hierarchy if the most-preferred plant species is not available."

For climatic reasons, neither *P. machaon* in Fennoscandia nor *P. zelicaon* in Washington State has the potential life-history flexibility associated with host selection by Californian *P. zelicaon*. Potentially variable phenology adds another layer of complexity to the selective forces which could be acting on host selection and ecotype formation.

The ecological differentiation of *P. zelicaon* populations may have been noticed by Lepidopterists long before, but it first received widespread attention when Charles L. Remington (1968) described a new species, *P. gothica*, from Colorado — a species hitherto subsumed under *P. zelicaon*. New sibling species of North American butterflies are described infrequently, and Remington's surprising claims stirred much interest and comment.

Papilio gothica was said to differ from Californian *P. zelicaon* in its phenology (univoltine vs. multivoltine) and host plant (*Pseudocymopterus montanus* [Gray] Coulter & Rose vs. *Foeniculum vulgare* Mill., both Umbelliferae) and in some very minor and statistical wing-pattern traits (the actual statistics were not given). The chief basis for alleging a difference at the species level lay in their hybrid phenotypes when each was crossed to tester stocks of *P. polyxenes* and *P. bairdii* Edw. This was a novel basis for such a claim, and seemed to have little foundation in the theory of speciation. Actual or potential reproductive isolation between *P. gothica* and *P. zelicaon* was not demonstrated or even tested for. Since no absolutely diagnostic character was identified for separating them, Remington's claim of species status for *P. gothica* failed to satisfy either the biological species concept or a morphological-typological one. Not unsurprisingly, it was severely criticized on genetic (Clarke & Sheppard, 1970), phenotypic (Shapiro, 1975) and ecological (Emmel & Shields, 1979) grounds. *P. gothica* is now treated as an ecotype or at best, a subspecies of *P. zelicaon*. Remington's claim, while not supported by most biologists, did serve to stimulate interest in the remarkable ecological versatility of *P. zelicaon* — and thus to trigger the whole cascade of parallel investigation discussed here.

Ironically, the genetic distinctness of Colorado *P. "gothica"* vis-à-vis other populations remains untested, despite the development of many sophisticated approaches since 1968. Ironically also, Remington's ideas about the genetics of speciation led to the findings by Charles Oliver (1972) that substantial reproductive incompatibility — approaching species status — could arise by sheer geographic separation (isolation-by-distance) within phenotypically undifferentiated morphospecies. Thus, even if substantial incompatibility were found between Californian *zelicaon* and *gothica* from the type locality, would this be sufficient to vindicate Remington's taxonomic judgement?

Papilio zelicaon occurs from the Golden Gate across the crest of the Sierra Nevada, some 290 km away. It is one of several butterfly morphospecies occurring at all ten stations on a permanent transect I have set up parallel to Interstate Highway 80, comprising tidal marsh, Inner Coast Range, Sacramento Valley riparian, west-slope Sierran foothill and montane forest, subalpine and alpine Sierran, and east-slope high desert and shrub/steppe environments (Fig. 1). All of the other ubiquitous species (*Pieris rapae* L., *Pontia protodice* Bdv. & LeC., *Colias eurytheme* Bdv., *Vanessa cardui* L., *V. annabella* Field, *Plebeius acmon* Westw. & Hew., etc.) are weedy, colonizing and altitudinally dispersing, and do not overwinter at the highest elevations (Fig. 2). Most have no "permanent" populations anywhere (Shapiro, 1980). Uniquely, *Papilio zelicaon* is not only a "permanent" resident at all ten stations (sampling periods ranging from 6-21 yr.), but shows four different life-history patterns along the transect, each of which translates into a genetically-based ecotype — and this scarcely exhausts the ecological versatility of *P. zelicaon* in the region!

THE DISTRIBUTION OF VOLTINISM

To be considered true ecotypes, ecologically different populations must be shown to reflect underlying genetic differences and not merely a plastic phenotypic response to environment. All of the voltinism patterns described below have a genetic basis under laboratory conditions, but they are not simple Mendelian phenomena. Diapause (developmental arrest, a specific form of dormancy) commonly behaves as a response to environmental conditions. The genetics of diapause in *P. zelicaon* is the quantitative genetics of a threshold character, complicated by the interactions between photoperiod and temperature and the roles of maternal effect and host plant on the diapause "decision" (Sims, 1980). Actually working out the genetics is complicated further by the long generation times when one is breeding from diapausing animals.

We believe the existing pattern of occurrence of voltinism reflects both ecological and historical forces, the latter often reflecting changes brought on by human beings. Before attempting to analyze how the pattern arose, a description of that pattern is in order.

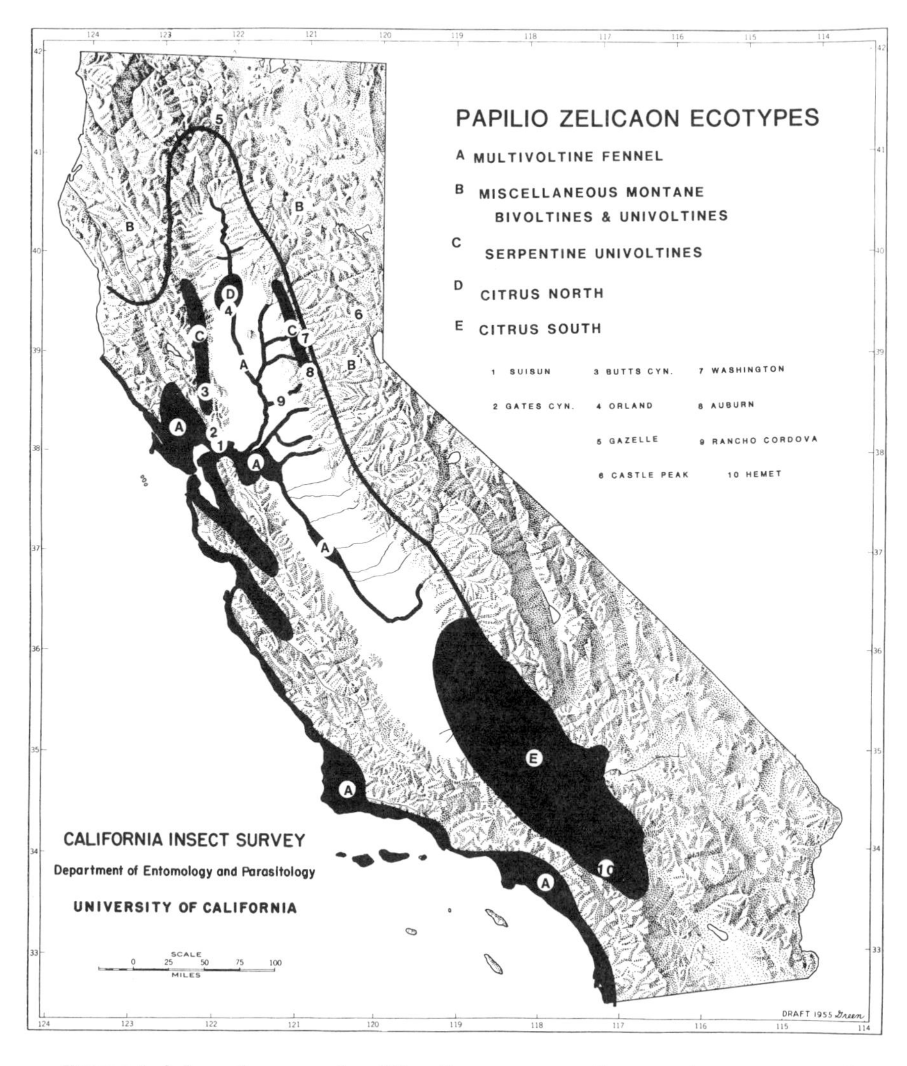

FIGURE 1. Schematic geography of *P. zelicaon* ecotypes. Ecotypes do not occur continuously within the shaded or delimited areas but rather as local populations in suitable patches of habitat. Alpine univoltines occur on high summits within the area of montane facultatively-bivoltine populations. Numbered populations were used in the electrophoretic analysis.

The most familiar populations of *P. zelicaon* — those taken by Remington as typical of the taxon — are strongly multivoltine, with 3-6 flights/year. Breeding may be effectively uninterrupted in coastal southern California, but even in San Diego County the capacity for facultative pupal diapause is retained. These populations occur in

FIGURE 2. *Papilio zelicaon* feeding at flowers of rabbitbrush, *Chrysothamnus nauseosus* (Pall.) Britton (Compositae) at Donner Pass, Nevada Co., 2100 m. This individual belongs to the rare second flight, which usually emerges too late to reproduce successfully at this elevation.

conjunction with sweet fennel, *Foeniculum vulgare*, which is their most important and often exclusive host. In some areas, poison hemlock, *Conium maculatum* L., is a significant host in the first half of the season. Multivoltine populations occur at elevations below 700 m (north) to 1200 m (south) in urban, suburban, ruderal and agricultural settings in and around the San Francisco Bay area, southern coastal plains, and Central (Sacramento and San Joaquin) Valleys in places which rarely experience hard freezes (Fig. 3). In the Central Valley they also occur in riparian (gallery) forest and in the coastal fog belt on hillsides where sweet fennel is now a component of grassland and coastal scrub. They follow roadsides and railways into the Inner Coast Range, Trinity and Sierra Nevada foothills, occurring locally around gold country towns such as Weaverville, Auburn, Colfax, Placerville, Jackson, Angel's Camp, and Sonora. There is some variation in the strength of multivoltinism among these populations, *e.g.*, Rancho Cordova animals are statistically more prone to diapause than North Sacramento ones (both Sacramento County).

Equally multivoltine populations which use cultivated citrus as an exclusive host occur in two widely-separated parts of the state: in the northern Sacramento Valley near Chico (Corning, Orland) and from the southern San Joaquin Valley south into Riverside County. Strikingly, casual infestation of ornamental or small-scale agricultural citrus is unreported elsewhere in the state, and I have never seen it occur in the Winters (Yolo Co.) "banana belt," despite careful search.

Strongly genetically univoltine populations occur on serpentine soils (Fig. 5) in the North Inner Coast Range in Napa, Colusa and Mendocino Counties from 300-1500 m+. They feed on native perennial umbellifers of the genus *Lomatium*. In laboratory culture they give 70-90+% diapause pupae in conditions which inhibit diapause almost completely in Valley stocks, and over 30% of the pupae require more than one year to break diapause.

Serpentine populations are fewer, smaller and more isolated on the Sierran west slope, but are known from at least Nevada, Placer and El Dorado Counties. They are less strongly univoltine than the Coast Range ones, perhaps because of more frequent gene flow from adjacent non-serpentine populations (see below); both in the lab and the field these populations may produce at least partial second broods some years.

Most montane *P. zelicaon*, from 500 m to about 2000 m in all the elevated regions of northern and central California, are mostly univoltine afield but facultatively produce a second brood — frequent in some localities but rare in others (Figs. 2, 6). Spontaneous second broods occur in about 20% of years at Donner Pass (2100 m), usually consisting of few individuals and too late for successful reproduction to occur before the hosts senesce or heavy snow and low temperatures intervene. Such emergences must be considered physiological "errors" which are selected against. In the northeast plateau (Fig. 7) and in moist valleys east of the Sierra (Sierra Valley, Carson Valley) such second broods may have a better chance of success, but are still risky due to the unpredictable onset of very cold

FIGURE 3. Habitat of multivoltine *Foeniculum*-feeding *P. zelicaon* in an urban vacant lot (Suisun City, Solano Co., California; sea level).

FIGURE 4. Fresh-water marsh habitat where a multivoltine ecotype of *P. zelicaon* might have existed in pre-American time (see text). Suisun Marsh, Solano Co., California.

FIGURE 5. Habitat of univoltine *P. zelicaon* on serpentine at Butts Canyon, Napa Co., California, in the North Coast Range. *Lomatium* abundant on talus in foreground.

FIGURE 6. Habitat of univoltine *P. zelicaon* at summit of Castle Peak, Nevada Co., California, in the northern Sierra Nevada (2700 m); an andesitic mudflow (lahar).

FIGURE 7. Habitat of partially bivoltine *P. zelicaon* near Gazelle, Siskiyou Co., California, in a shrub/steppe – grassland environment.

nights. They are also fairly frequent in northwestern California (Trinity and western Shasta and parts of Mendocino Counties).

These facultatively bivoltine, perhaps incipiently multivoltine populations are apparently heterogeneous both ecologically and genetically. Their host plants vary locally, with the large native Umbellifer *Angelica* perhaps the most frequent choice. In northeastern and east-central California, poison hemlock is frequently used. The physiological responses of these populations fall in between those of serpentine univoltines and valley multivoltines. In general, west-slope Sierran montane populations tested have a lower propensity to direct development than east-slope and summit ones.

Subalpine and alpine *zelicaon* have no choice but to be univoltine, given their short seasons. In the laboratory, however, they appear no different from nearby montane populations in propensity to develop without diapause. Because *P. zelicaon* is a hilltopping species (Shields, 1967) and males may travel several miles daily in rugged topography in search of mates, gene flow should be extensive and would be predicted to prevent differentiation of an alpine ecotype (as was argued for *Pontia occidentalis* Reakirt by Shapiro, 1974). The facts support this interpretation, even though the hosts at and near tree-line are different (often *Cymopterus terebinthinus* [Hook] T. & G.) from those used below (*e.g.*, *Angelica*). It is noteworthy that both the tendency to diapause and the occurrence of multiple-year diapause are significantly higher in Coast Range serpentine univoltines than in tree-line Sierran populations from 2500m higher.

Foeniculum vulgare has spread aggressively into foothill areas during the 20 years we have been interested in *P. zelicaon*. It has been tracked closely by multivoltine *zelicaon*, which are now widespread in foothill areas where they were unknown only a few years ago. In some cases the spread of sweet fennel has brought multivoltine *zelicaon* into close proximity to serpentine (or non-serpentine) univoltine populations, setting up potentially very interesting gene-flow situations. All of the ecotypes are reciprocally intercompatible in the laboratory, except for an occasional male-biased sex ratio in some crosses between serpentine univoltines and other ecotypes. The strong selection pressures expected to act locally, depending on the dispersion of ephemeral *vs.* persistent hosts, could set up a situation reminiscent of the rapid buildup of isolation in some cases of heavy-metal tolerance in plants on mine tailings (Bradshaw, 1971).

Sims (1980) made the beginnings of a quantitative-genetic study of voltinism in *P. zelicaon*. Subsequent rearings have confirmed the ready selectability of diapause characteristics in mass-selection experiments (Shapiro, unpublished data). At the same time, the diversity of genetic systems revealed by hybridizing different montane, facultatively bivoltine populations against standard tester stocks (Napa County serpentine univoltines, Suisun Marsh multivoltines) suggests that similar phenologies have been arrived at in different ways in different places. Just as in the case of polytopic subspecies in taxonomy, it is not immediately apparent whether it will be possible to distinguish uniquely derived traits spread through biogeographic processes from traits repeatedly derived in parallel in different locations. The montane *zelicaon* do, however, appear genetically heterogeneous even if phenologically similar (Shapiro, unpublished data).

HISTORICAL DERIVATION OF ECOTYPES

The *machaon* group is recorded on many genera of Umbelliferae. The native umbellifer flora of California is large (30 genera, of which at least 21 are edible to *Papilio*: Munz & Keck, 1970). Only three native rutaceous genera occur in California, and only one of those (*Ptelea crenulata* Greene) is sympatric with *P. zelicaon* in the northern half of the state. *Ptelea* is a known host of *Papilio cresphontes* Cramer, an exclusive Rutaceous feeder, elsewhere in North America, but *P. zelicaon* has not been recorded on it, and I know of only one circumstantial record of association in the state.

Most of the native Umbelliferae of low to moderate elevations in Mediterranean California climates are strongly seasonal perennials which remain green from two to four or five months. Emmel & Shields (1979) noted that such species as *Lomatium dasycarpum* (T. & G.) Math. & Const., *L. californicum* (Nutt.) Math. & Const., *L. marginatum* (Benth.) C. & R., and *L. utriculatum* (Nutt.) C. & R. in the North Coast Ranges could not normally sustain more than one generation of *P. zelicaon* per year, and no alternative hosts are available for a second. For others, such as *Tauschia parishii* (C. & R.) Macbr., *Oenanthe sarmentosa* Presl., *Heracleum sphondylium* L., or some of the large foothill *Angelica,* "there is a small percentage of pupae which do not diapause and emerge as a second brood," with enough plant material still available to permit successful reproduction at least in some years.

All of the multivoltine populations known in California are, as previously noted, associated with either *Foeniculum vulgare* or citrus. Bivoltine populations are often associated with another naturalized weed, *Conium maculatum*, and multivoltine populations may use it to some extent during its seasonal period of availability. All of these plants come from the Old World as byproducts of human activity, raising the question: where did the multivoltine ecotype(s) come from, how and how long ago?

The date of introduction of sweet fennel into California is unknown. Frenkel (1970) follows Robbins (1940) in attributing it to the period of American settlement (1849-1860) based not on positive evidence but on the plant's absence from adobe bricks of Mission Period provenance. Most Old World weeds that came into California around the Gold Rush had already been established in Chile for a century or more and came here as seeds in the fur of domestic animals or in fodder for them, loaded on ships at Valparaiso or Concepción. Sweet fennel is ubiquitous in Chile, but to the naked eye is varietally distinguishable from our plant. My group has chosen to treat the plant as a probable, but unproven, Mission Period introduction. This is the "conservative" option insofar as it maximizes the time available for the evolution of fennel feeding. Emmel & Shields (1979) agree. Discussing Remington's assumption that nominate Californian *zelicaon* is multivoltine, they observe: "At the time the type was collected (ca. 1850), the extent of introduced *Foeniculum vulgare* may have been very small indeed, and *Citrus* was not being grown on a large scale. The type, even if taken from a lowland area, could have been from a univoltine population. The development of the multivoltine, *Foeniculum*-feeding and *Citrus*-feeding ecotypes probably took place toward the end of the 19th Century as these two foodplants became increasingly abundant." The only biological information published on *P. zelicaon* before 1880 appears in Boisduval's *Lepidoptères de la Californie* (1868). In Part I, p. 11, he says: "Il paraît aux mêmes époques que notre *machaon*." ("It appears at the same times as our *machaon*.") In Part II, p. 37: "M. Lorquin a souvent élevé la chenille du papillon *Zolicaon* que nous avons décrit il y a bientôt dix-sept ans. Elle n'est pas très rare dans la campagne sur différentes ombellifères, et même quelquefois, sur le fenouil et les carottes cultivés dans les jardins . . . éclot en mai et en août." ("Mr. Lorquin has often raised the caterpillar of the butterfly *Zolicaon* which we described seventeen years ago. It is not very rare in the countryside on different umbellifers, and even sometimes on fennel and carrots cultivated in gardens . . . ecloses in May and August.") The location of this observation is not given, but it indicates bivoltinism (May and August, a pattern seen rarely today and mostly in bivoltine or partially bivoltine montane or foothill populations). Lorquin might have missed an earlier brood (March?). There is no mention of citrus.

Citrus sinensis Osbeck was introduced to California by the Franciscan fathers in the Mission Period. The first commercially profitable orchard was in operation in Los Angeles in 1841 (Opitz & Platt, 1969). The history of the orange is spelled out in some detail by Hittell (1866). At the time of writing (1862), "There are now, so far as I can learn, about 2500 orange trees set out in orchard in the state, more than 2/3 of them being in the orchard of William Wolfskill, in the town of Los Angeles. About 400 of the orange-trees in the state are old — from ten to fifty years of age . . . (Some of the original trees planted by the Franciscans at San Diego, San Fernando, San Juan Capistrano, *etc*.) are still standing as monuments to the industry and enterprise of the old priests." The citrus industry was given further impetus with the introduction of the Washington navel orange from Brazil (1873) and the Valencia from England (1876). In 1975 some 91,000 ha were planted in citrus in California and Arizona.

The first report of *P. zelicaon* on citrus is by Coolidge (1910, based on 1909 data from Tulare County). He attributed the apparent switching of host plants to the encroachment of civilization into undisturbed, native habitats and the declining abundance of native hosts. Horton (1922) gave a detailed account of damage to southern California orchards by *P. zelicaon*. The first notice of damage to citrus in the northern Sacramento Valley dates from the late 1960s to early 1970s (P. Meith to K. Masuda, pers. commun.).

All authors who have commented on the origin of the multivoltine fennel and citrus ecotypes of *P. zelicaon* (Emmel and Shields, Sims, Masuda, Tong) agree that they probably originated since the Mission Period, since no hosts were available at low elevations in Mediterranean California to sustain summer breeding before that time. These discussions imply, but never explicitly claim, that the multivoltine ecotypes each arose once and subsequently spread, following the introduced host. Various lines of evidence cast reasonable doubt on that inference.

As noted previously, populations of *P. zelicaon* studied by Thompson in Washington feed naturally on native perennial Umbelliferae and have no prior exposure to sweet fennel (which in that state is confined to disturbed urban and suburban environments where severe freezes are uncommon). When sweet fennel was included as an option in Thompson's preference tests, it proved highly acceptable; indeed, the Composite-feeding *P. oregonius* accepted it as well even in a free-choice situation. This seems to imply that fennel would be discovered quickly wherever it appeared in *P. zelicaon* country. Thompson comments (1988a): "Some females of both *Papilio* species laid eggs on the novel host, *F. vulgare,* suggesting that these species could shift onto this plant species in areas in which their preferred hosts were scarce or absent . . . In fact, *P. zelicaon* populations along the western coast of North America from British Columbia to California now feed regularly on introduced fennel . . . and one population in downtown Seattle seems to feed exclusively on fennel (R. M. Pyle, pers. commun.; Thompson, pers, observ.)."

Boisduval's data are presumably based on correspondence from

Lorquin prior to 1857, when Lorquin took ship for southeast Asia. He had collected from Mount Shasta to Los Angeles, and there is no available basis for localizing his very early fennel record. However, it is not necessary to do so. The standard scenario envisions uni- or weakly bivoltine populations of *P. zelicaon* encountering fennel, beginning to use it, and undergoing rapid selection for shortened generation time through the weakening of diapause. Sims (1980) demonstrated experimentally that diapause was a highly selectable trait in the laboratory. He also showed that what is selected is not the ability to diapause *per se*, but the thresholds governing induction and termination of diapause; even populations which do not usually diapause have the potential to do so, and populations which normally diapause may develop continuously in exceptional circumstances (Shapiro, 1984). This process could have occurred repeatedly as more and more populations were exposed to fennel. Alternatively, gene flow outward from fennel-using populations could have hastened local adaptation to fennel elsewhere, a situation discussed further elsewhere in this paper.

A careful review of the native umbellifer flora suggests a *caveat* to this scenario: a truly multivoltine ecotype could have existed in pre-Spanish California in one special habitat, now nearly extinct. In the Central Valley there were very extensive seasonal wetlands or *tulares* (Bakker, 1971; Sculley, 1973; Thompson, 1961) formed by the impoundment of stream drainage by the natural levees of the larger rivers, especially the Sacramento. These marshes were covered by "tules, cattails, bulrushes, sedges and various aquatic herbs, with intervening areas of shallow open water which dried during late summer" (Sculley, 1973). Attempts to drain the marshes began in the mid 1860s and proceeded piecemeal until the formation of reclamation districts in the early 1900s. Work was completed on the Fremont Weir and Yolo Bypass about 1915, effectively eradicating the Sacramento marshes. No detailed floristic inventory exists of these marshes, but the Suisun Marsh (Solano County) is the largest reasonably intact remnant of them, and some idea of their original flora can be obtained from it. Three common and widespread Umbelliferae remain green in the marsh well into summer: Water hemlock, *Cicuta Bolanderi* Wats. and *C. Douglasii* (DC.) Coult. & Rose; *Oenanthe sarmentosa* Presl.; and celery, *Apium graveolens* L. All of these could sustain breeding at least into August. Celery is naturalized from

Europe (dated to the American period by Frenkel), but the other two are native. Although the Suisun population now feeds on fennel, I have found eggs and larvae occasionally on *Cicuta,* and experimentally it is a very good host in the Sierra Nevada (M. Nelson, unpublished data). We do not think of *P. zelicaon* as a marsh butterfly today, but the British subspecies of *P. machaon* is restricted to marshes and uses a single umbelliferous host there (Dennis, 1977). If a multivoltine marsh ecotype existed historically, it would have been preadapted to the seasonality of both fennel and citrus (Fig. 4).

ORIGIN OF ECOTYPES: THE NORTHERN CITRUS CASE

The ecology of the northern Sacramento Valley citrus ecotype of *P. zelicaon* was studied in depth by K. Masuda (1981), who was interested in the conditions under which host shifts occur in phytophagous insects. The *machaon* group had been the focus of previous work on host selection and host shifts, centering on the chemical basis of oviposition and larval feeding and the applicability of the Hopkins Host Selection Principle (Hopkins, 1917) to preferences in the group. This notion — that adult females will prefer to oviposit on the plant they had eaten as larvae — has been advanced to account for the origins of "phytophagous varieties" and sibling species and even as a basis for "sympatric speciation." It has not been found operative in the *machaon* group, including *zelicaon* as studied by Masuda. (If it were operative, this would further reinforce the development of ecotypes.)

Wiklund (1974) investigated the mechanism of host selection by Swedish *P. machaon.* He reared the progeny of a female from Vejbystrand, where the normal host is *Angelica archangelica* Thell., on four umbelliferous species, finding that the most-preferred host was *Peucedanum palustre* Moench., a species which does not occur at Vejbystrand but which is the sole British host (Dennis, 1977). *Peucedanum* was an intrinsically desirable host, producing rapid growth and high potential fecundity; that is, the preference hierarchy coincided overall with suitability for larval growth, regardless of individual or local population experience. The Vejbystrand population would presumably have switched to *Peucedanum,* given the chance. Was the switch to citrus in California mediated in the same way?

Masuda tested the oviposition preferences of *Foeniculum* and citrus-feeding multivoltine *zelicaon* from the Sacramento Valley. A symmetrical design was employed, in which each ecotype was reared in split-brood fashion on both hosts and mated females of each tested on both. However, survivorship on citrus was so poor that ultimately wild citrus females (with experience of citrus but not fennel) had to be used. (No fennel was found within 8 km of the collection site.) The complete protocols can be found in Masuda (1981). Oviposition preferences were tested using a randomized complete-block array of size-matched cuttings (Little & Hills 1975) including an unattractive control (the landscape shrub *Escallonia,* Escalloniaceae). The results (Table 1) are consistent across all groups and significant in all but the citrus/fennel one: fennel was preferred over citrus.

When the preferences of the individual females are examined (Table 2), a range of preferences can be found, with some individuals preferring one host or the other to varying degrees. There was no support for the Hopkins Host-Selection Principle, but if the individual differences had a genetic component (this was not pursued further), the opportunity was present for selection to act on oviposition preference, as it has apparently done in some cases described later in this paper.

Since Wiklund found evidence that the preference hierarchy was correlated with performance on the various hosts (even when a given host was new to the population tested), Masuda assessed the growth and fecundity of *P. zelicaon* on both citrus and fennel (Tables 3-8, Figs. 8-9). The full protocols again can be found in Masuda (1981). The results of these studies can be summarized as follows for the citrus ecotype:

1. Mean duration of larval development was longer on citrus than on fennel.

TABLE 1. Ovipositional preferences of *P. zelicaon* in a laboratory choice situation. Significance refers to Student's t-test applied to mean numbers of eggs laid on *Foeniculum* vs. *Citrus*.

Group	*Host Plant*		*Sample*	*Mean Number of Eggs/Host Plant*			*Significance*
Number	*Parental*	*Larval*	*Size*	*Foeniculum*	*Citrus*	*Escallonia*	
I	*Citrus*	*Citrus*	20	69.7	24.0	3.2	$p < 0.05$
II	*Citrus*	*Foeniculum*	10	68.5	10.6	2.9	$p < 0.05$
III	*Foeniculum*	*Foeniculum*	23	185.5	15.7	1.5	$p < 0.05$
IV	*Foeniculum*	*Citrus*	9	77.9	47.7	4.7	ns

TABLE 2. An example of individual variation in oviposition preference among females. These data from *Citrus*-reared females from *Foeniculum*-feeding populations.

ID[1]	*F*	*C*	*E*
C169	70	44	1
C173	263	48	5
C186	83	142	1
C320	24	10	1
C321	16	1	1
C301	13	15	3
C306	20	131	19
C304	160	37	8
C322	52	1	3

[1] Symbols: ID = identification number of female; F = *Foeniculum*; C = *Citrus*; E = *Escallonia*

2. Larval survivorship was lower on citrus than on fennel.
3. Pupal weight, which is positively and significantly correlated with subsequent female fecundity, was lower on citrus than on fennel.
4. There were no significant host plant effects on the occurrence of diapause.
5. There were no significant host plant effects on pupal survivorship.

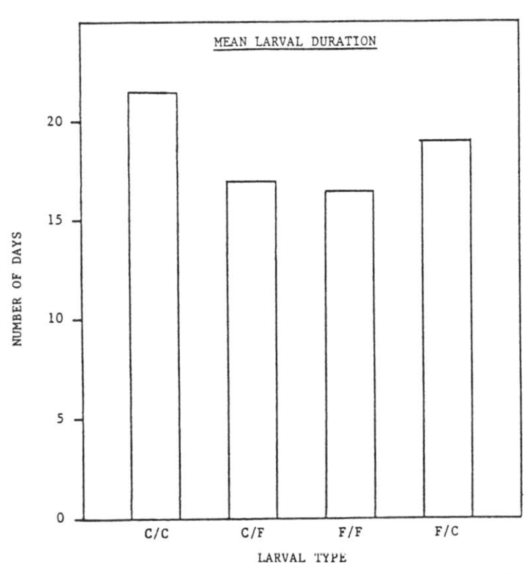

FIGURE 8. Mean larval duration of groups of *P. zelicaon* reared under uniform conditions but with different host plant histories. The first letter is the ecotype; the second the actual plant used. F = *Foeniculum*; C = *Citrus*.

HOST PLANT: LARVAL	PARENTAL	Citrus / Citrus	Citrus / Foeniculum	Foeniculum / Foeniculum	Foeniculum / Citrus
Citrus	Citrus	—	t'= 4.56 t = 2.19 *	t'= 5.13 t = 2.19 *	t'= 2.88 t = 2.18 *
Citrus	Foeniculum	t'= 4.56 t = 2.19 *	—	t'= 1.80 t = 1.98 ns	t'= 4.50 t = 2.00 *
Foeniculum	Foeniculum	t'= 5.13 t = 2.19 *	t'= 1.80 t = 1.98 ns	—	t'= 6.45 t = 1.99 *
Foeniculum	Citrus	t'= 2.88 t = 2.18 *	t'= 4.50 t = 2.00 *	t'= 6.45 t = 1.99 *	—

TABLE 3. Comparison of larval duration among groups of *P. zelicaon* reared under uniform conditions but with different host plant histories. t' = calculated t-values adjusted for unequal sample sizes, t = t-values at $p = 0.05$ adjusted for unequal sample sizes. Values above and below the diagonal are identical.

HOST PLANT: LARVAL	PARENTAL	Citrus / Citrus	Citrus / Foeniculum	Foeniculum / Foeniculum	Foeniculum / Citrus
Citrus	Citrus	—	X^2= 71.2 *	X^2= 125.4 *	X^2= 39.6 *
Citrus	Foeniculum	X^2= 71.2 *	—	X^2= 0.14 ns	—
Foeniculum	Foeniculum	X^2= 125.4 *	X^2= 0.14 ns	—	X^2= 11.5 *
Foeniculum	Citrus	X^2= 39.6 *	—	X^2= 11.5 *	—

TABLE 4. Comparison of larval survivorship among groups of *P. zelicaon* with different host plant histories. Chi square values with df = 1 are given; significance level $p \leq 0.05$. Values above and below the diagonal are identical.

HOST PLANT: LARVAL	HOST PLANT: PARENTAL	Larval: Citrus / Parental: Citrus	Larval: Citrus / Parental: Foeniculum	Larval: Foeniculum / Parental: Foeniculum	Larval: Foeniculum / Parental: Citrus
Citrus	Citrus		x^2= 0.04 ns	x^2= 0.4 ns	x^2= 0.1 ns
Citrus	Foeniculum	x^2= 0.04 ns		x^2= 0.4 ns	
Foeniculum	Foeniculum	x^2= 0.4 ns	x^2= 0.4 ns		x^2= 2.3 ns
Foeniculum	Citrus	x^2= 0.1 ns		x^2= 2.3 ns	

TABLE 5. Comparison of diapause incidence among pupae of *P. zelicaon* with different host plant histories. Chi-square values with df = 1. Values above and below the diagonal are identical.

HOST PLANT: LARVAL	HOST PLANT: PARENTAL	Larval: Citrus / Parental: Citrus	Larval: Citrus / Parental: Foeniculum	Larval: Foeniculum / Parental: Foeniculum	Larval: Foeniculum / Parental: Citrus
Citrus	Citrus		x^2= 0.02 ns	x^2= 0.06 ns	x^2= 0.03 ns
Citrus	Foeniculum	x^2= 0.02 ns		x^2= 0.03 ns	
Foeniculum	Foeniculum	x^2= 0.06 ns	x^2= 0.03 ns		x^2= 0.07 ns
Foeniculum	Citrus	x^2= 0.03 ns		x^2= 0.07 ns	

TABLE 6. Comparison of pupal mortality among *P. zelicaon* with different host plant histories. Chi-square values with df = 1. Values above and below the diagonal are identical.

HOST PLANT LARVAL / PARENTAL	LARVAL: Citrus PARENTAL: Citrus	LARVAL: Citrus PARENTAL: Foeniculum	LARVAL: Foeniculum PARENTAL: Foeniculum	LARVAL: Foeniculum PARENTAL: Citrus
Citrus / Citrus	X	t'= 4.5020 t = 2.1480 *	t'= 9.1842 t = 2.1251 *	t'= 1.2027 t = 2.3151 ns
Citrus / Foeniculum	t'= 4.5020 t = 2.1480 *	X	t'= 6.6503 t = 2.0822 *	t'= 1.6925 t = 2.3278 ns
Foeniculum / Foeniculum	t'= 9.1842 t = 2.1251 *	t'= 6.6503 t = 2.0822 *	X	t'= 5.8399 t = 2.2716 *
Foeniculum / Citrus	t'= 1.2027 t = 2.3151 ns	t'= 1.6925 t = 2.3278 ns	t'= 5.8399 t = 2.2716 *	X

TABLE 7. Comparison of pupal weight among *P. zelicaon* with different host plant histories. Statistical procedures as in Table 3.

HOST PLANT (LARVAL / PARENTAL)	Citrus / Citrus	Citrus / Foeniculum	Foeniculum / Foeniculum	Foeniculum / Citrus
Larval Survivorship	12.0 % n= 108	70.3 % n= 101	72.8 % n= 368	54.2 % n= 96
Pupal Mortality	0 % n= 13	5.6 % n= 71	6.0 % n= 268	3.8 % n= 52
Pupal Diapause	30.8 % n= 13	38.0 % n= 71	43.3 % n= 268	30.8 % n= 52

TABLE 8. Comparison of life-history parameters of *P. zelicaon* with different host plant histories.

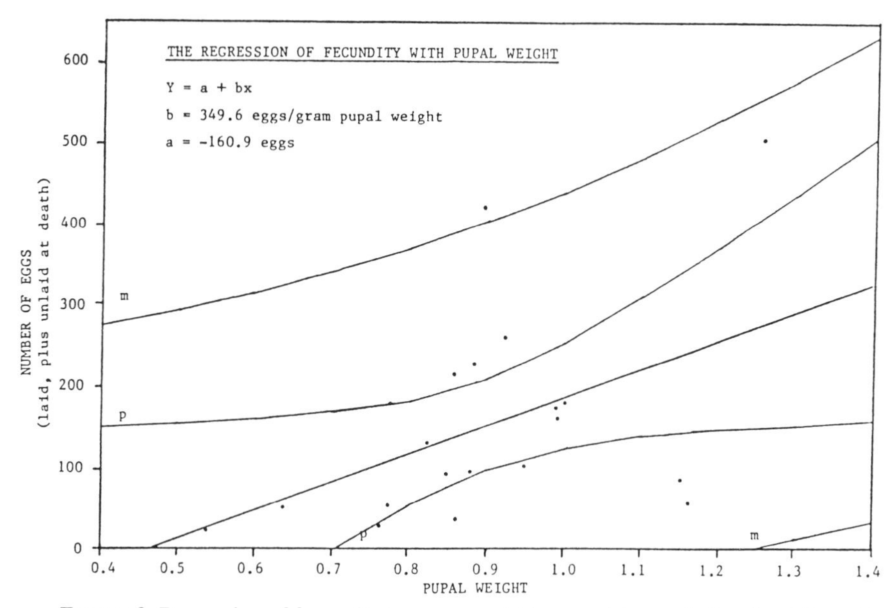

FIGURE 9. Regression of fecundity (mean eggs laid + remaining unlaid at death/female) on pupal weight, with 95% confidence intervals for the mean (m) and population (p) regression lines. This analysis permits pupal weight to be used as a surrogate for fecundity.

In summary, these studies provide no evidence that citrus is an intrinsically attractive or desirable host in comparison with fennel; in fact, it is less desirable on three of five scales, and no better on the others. Masuda also conducted a "trap-cropping" experiment in which potted fennel and citrus plants were set in citrus orchards and in vacant lots full of fennel (at Orland and Fairfield, CA, respectively). Results are given in Table 9. No conclusions could be drawn at Fairfield because so few eggs were laid, but at Orland potted fennel received many eggs. This is an especially striking result since the growth form of fennel is so different from citrus, and wild females had no experience with fennel in their lifetimes and were accustomed to flying high in the trees in order to oviposit on tender shoot tips.

TABLE 9. Results of field oviposition tests for *P. zelicaon* in the habitats of multivoltine *Foeniculum* and *Citrus* ecotypes ($\bar{x} \pm$ SD)

		Mean (± SD) *Number of Eggs per Host Plant*		
Environmental Setting	*Sample Size*	*F. vulgare*	*C. sinensis*	*E. exoniensis*
Citrus orchard Orland, Glenn Co.	10	6.00 (± 3.90)	0.10 (± 0.30)	0
Vacant lot Fairfield, Solano Co.	12	0.08 (± 0.29)	0	0

Masuda concluded that there was enough evidence of genetically-based differences to justify recognition of a citrus ecotype, but that its origins were best explained in terms of expansion into a previously vacant niche — *i.e.*, it is better to utilize a poor-quality host in an area where no other host is available, than to be excluded from that area altogether. The behavioral signature of the citrus ecotype is not a *preference* for citrus, but rather a willingness to oviposit on it *at all.* (By contrast, Masuda's results reinforce the impression that fennel is an intrinsically desirable host.)

One alternative left unexplored by Masuda was that the Sacramento Valley citrus ecotype did not arise *in situ* from nearby multivoltine fennel feeders, but rather was introduced from southern California on nursery stock. We approached this genetically.

HOW DIFFERENT ARE THE ECOTYPES GENOMICALLY?

The electrophoretic investigation by Tong & Shapiro (1989) began as a test of the hypothesis that the northern citrus ecotype was introduced from the south, but quickly expanded to incorporate much of the ecotypic diversity available to us in northern California. Fourteen loci were surveyed in ten populations, representing the northeast plateau univoltines, montane partial bivoltines, alpine, serpentine univoltine, fennel and citrus multivoltine life histories (Table 10); wild-collected sample sizes varied from 6-35. Only three loci were polymorphic (Table 11; the full protocols appear in Tong & Shapiro, 1989). Because the populations were all extremely similar (Table 11-12), an identity matrix and dendrogram were prepared using only data from the polymorphic loci (Table 13, Fig. 10) to maximize the discriminatory content of the data set. Because of the low variability and the procedures used, not too much should be read into the resulting pattern — there is too much potential error. The most parsimonious interpretation is that the northern citrus ecotype is locally derived, not introduced from the south, but the data do not allow us to exclude this possibility.

The most interesting point, however, is the extreme similarity of all the populations (except Gazelle), despite their vast differences in ecology and especially phenology-related physiology. Since we know most of the latter is genetically-determined, we conclude that ecotype formation in Californian *P. zelicaon* has involved strong local selection on life history but no large-scale genomic reorganiza-

TABLE 10. Populations used in electrophoretic comparisons of *P. zelicaon* ecotypes

Population	*Voltinism*	*Elevation*[1] (m)	*Habitat*	*Hosts*
Suisun Marsh, Solano Co.	M	1	levees, ruderal, marsh	*Foeniculum*
Gates Canyon[2], Solano Co. (Coast Range)	M	60-762	foothill riparian	*Foeniculum*
Butts Canyon, Napa Co. (Coast Range)	U	457	serpentine chaparral	*Lomatium*
Orland area, Butte Co. (Sacramento Valley)	M	67	orchards	*Citrus*
Gazelle, Siskiyou Co.	B	838	grassland shrub/steppe	*Angelica, Conium*
Castle Peak, Nevada Co. (Sierra Nevada)	U	2743	alpine fell-field	*Cymopterus*
Washington, Nevada Co. (Sierra Nevada)	U	1220	serpentine chaparral	*Lomatium*
Auburn, Placer Co. (Sierra Nevada)	U/M[3]	366	canyon riparian, ruderal	*Foeniculum,* natives
Rancho Cordova, Sacramento Co. (Sacramento Valley)	M	9	riparian, dredge tailings	*Foeniculum, Conium*
Hemet, Riverside Co.	M	487	orchards	*Citrus*

[1] M = multivoltine; B = bivoltine; U = univoltine
[2] Referred to as "Blue Ridge" in data tables from Tong & Shapiro (1989).
[3] Auburn is a contact point for foothill univoltine and fennel-multivoltine populations.

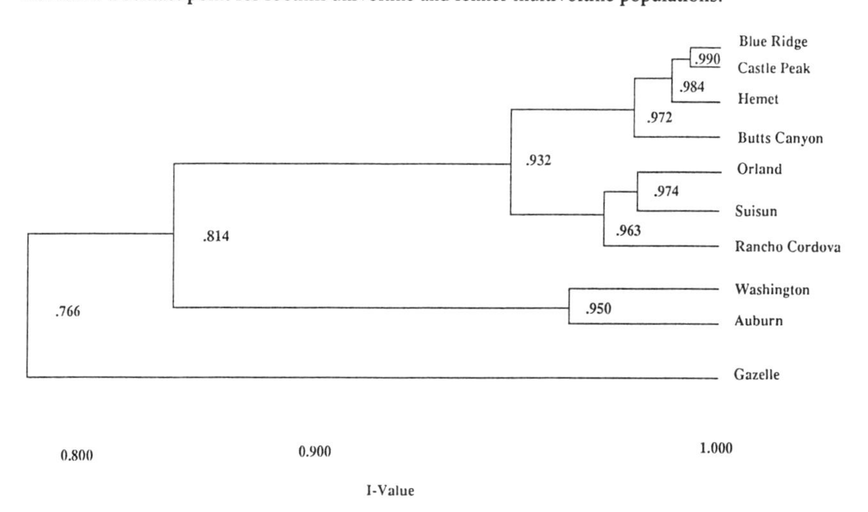

FIGURE 10. Dendrogram illustrating genetic identity relationships for all populations, using only the three polymorphic loci (see text and Table 13).

TABLE 11. Electromorph frequencies for all loci. (For complete protocols see Tong & Shapiro, 1989).

Locus	*Allele*	*Gazelle*	*Washington*	*Hemet*	*Orland*	*Butts Canyon*	*Suisun*	*Blue Ridge*	*Rancho Cordova*	*Castle Peak*	*Auburn*	*Overall*
	n	*8*	*6*	*19*	*19*	*14*	*24*	*7*	*16*	*35*	*8*	*157*
PGI	13	0	0	0	.05	.20	0	0	.111	.128	0	140
	7	1.0	1.0	1.0	.95	.80	1.0	1.0	.889	.872	1.0	.859
ALDO	10	1.0	1.0	1.0	1.0	1.0	1.0	1.0	1.0	1.0	1.0	1.0
aGPD	23	1.0	1.0	1.0	1.0	1.0	1.0	1.0	1.0	1.0	1.0	1.0
GOT-1	27	1.0	1.0	1.0	1.0	1.0	1.0	1.0	1.0	1.0	1.0	1.0
HK-1	40	1.0	1.0	1.0	1.0	1.0	1.0	1.0	1.0	1.0	1.0	1.0
PGM	32	.50	0	.26	.10	.18	.23	.25	.29	.23	.15	.221
	26	.50	.20	.21	.47	.18	.46	.25	.35	.28	.23	.326
	23	0	.40	.32	.40	.45	.18	.33	.23	.35	.46	.311
	20	0	.40	.21	.03	.18	.14	.17	.03	.13	.15	.130
FUM-1	12	1.0	1.0	1.0	1.0	1.0	1.0	1.0	1.0	1.0	1.0	1.0
MPI	44	0	0	0	.06	0	.09	0	0	.02	0	.026
	40	0	0	0	.06	0	.09	0	0	.08	0	.039
	37	0	0	.30	.20	.43	.18	.40	.09	.30	0	.221
	33	.33	.50	.57	.60	.50	.64	.60	.73	.52	.40	.561
	28	.67	.50	.13	.06	.07	0	0	.18	.08	.60	.149
ME	20	1.0	1.0	1.0	1.0	1.0	1.0	1.0	1.0	1.0	1.0	1.0
G3PD	7	1.0	1.0	1.0	1.0	1.0	1.0	1.0	1.0	1.0	1.0	1.0
G6PD	7	1.0	1.0	1.0	1.0	1.0	1.0	1.0	1.0	1.0	1.0	1.0
HBDH	6	1.0	1.0	1.0	1.0	1.0	1.0	1.0	1.0	1.0	1.0	1.0
EST-1	22	1.0	1.0	1.0	1.0	1.0	1.0	1.0	1.0	1.0	1.0	1.0
EST-2	14	1.0	1.0	1.0	1.0	1.0	1.0	1.0	1.0	1.0	1.0	1.0

tion. The slight sex-ratio disruption occasionally seen in crosses involving serpentine univoltines is probably an epiphenomenon of diapause control: a deficit of the heterogametic sex (female in butterflies) in hybrid crosses involving diapause differences is caused by failure of the female pupae to break diapause spontaneously (*e.g.*, Bowden, 1953, 1957; such pupae can sometimes be rescued with allochthonous molting hormone). Because the geography of Californian *zelicaon* suggests a Quaternary origin for the ecotypes, the potential for further differentiation cannot be discounted.

GENE FLOW AND SELECTION: EQUILIBRIUM?

As noted earlier, multivoltine populations have appeared in foothill areas as sweet fennel becomes established in those areas. The most interesting situation directly observed to date is in the Vaca Hills, Inner Coast Range, west of Vacaville, Solano Co. When studies were

TABLE 12. I-value matrix for all populations using all 14 loci sampled.

	Gazelle	*Wash.*	*Hemet*	*Orland*	*Butts C.*	*Suisun*	*Blue R.*	*R. Cordova*	*C. Peak*	*Auburn*
Gazelle	–	.97	.96	.96	.95	.96	.96	.98	.97	.98
Wash.		–	.98	.97	.98	.97	.98	.98	.98	.99
Hemet			–	.99	.99	.99	.99	.99	.98	.97
Orland				–	.99	.99	.99	.99	.98	.97
Butts C.					–	.99	.99	.99	.99	.97
Suisun						–	.98	.99	.98	.96
Blue R.							–	.99	.99	.97
R. Cordova								–	.99	.98
C. Peak									–	.98
Auburn										–

TABLE 13. I-value matrix for all populations using only the 3 polymorphic loci.

	Gazelle	*Wash.*	*Hemet*	*Orland*	*Butts C.*	*Suisun*	*Blue R.*	*R. Cordova*	*C. Peak*	*Auburn*
Gazelle	–	.765	.713	.707	.656	.723	.676	.807	.730	.853
Wash.		–	.855	.830	.842	.795	.812	.849	.850	.950
Hemet			–	.953	.962	.960	.982	.953	.985	.784
Orland				–	.930	.974	.953	.957	.973	.783
Butts C.					–	.900	.984	.901	.979	.808
Suisun						–	.952	.966	.964	.719
Blue R.							–	.942	.990	.754
R. Cordova								–	.757	.822
C. Peak									–	.808
Auburn										–

initiated in this area in 1972 there was no fennel in the canyons, though a few plants did occur along the north-south Pleasants Valley into which those canyons open, and on the slopes facing Interstate 80 which traverses the range at a low pass south of them. Fennel was abundant at Lagoon Valley, where multivoltine *zelicaon* probably continuous with the very large Fairfield-Suisun populations occurred. Strays of this population were occasionally seen in Pleasants Valley, which is grazed annual grassland and thinned foothill oak woodland. In the Vaca Hills themselves only univoltine populations were observed. Males hilltopped on the ridgetops, and in the canyons occasional females and larvae could be found, utilizing native perennial umbellifers (in order of abundance, *Lomatium californicum* [Nutt.] Matt. & Const., *Angelica tomentosa* Wats., and *Perideridia Kelloggii* [Gray] Math.) which grow in mesic riparian woodland. As of 1972, a single patch of ten fennel plants existed on ranch land at the base of Gates Canyon, and it was uncolonized by *P. zelicaon.* The first oviposition seen on this patch was in 1975, when it was much more extensive. By 1978, fennel was spreading aggressively up Gates Canyon on the immediate roadsides, and the dates of captures or sightings suggested four generations there that year. As of 1992, there were over 800 fennel plants in Gates Canyon along several km of roadside, larvae are readily found all season, and adults can be seen from February or March to October, though less common in the second half of the season. On nearby ridgetops to the northwest, however, hilltoppers are seldom seen after May.

It was initially assumed that the multivoltine population at Lagoon Valley and along I-80 had merely spread up Pleasants Valley and thence into the canyons as fennel moved upslope. Electrophoretically, however, Tong's results indicate that the Gates Canyon population is genomically close to the serpentine univoltines from Napa County, not the Suisun multivoltines. There are two possibilities: either an existing univoltine population adapted very rapidly to fennel phenology, undergoing selection for weaker diapause *in situ*; or genes for multivoltinism introgressed in via occasional contact with the Lagoon Valley population, with little impact resulting at other loci. Before fennel entered the area there would have been strong selection against direct development; once it was there, such genes could have been selected for very rapidly.

Recent rearings from small, isolated serpentine populations in the western Sierra foothills have produced so many unexpected non-diapausers in the lab that the possibility exists of massive gene flow overwhelming local adaptation in these areas — particularly in the lower American River canyons, where there has been an explosive increase in the amount of fennel in the past decade.

Such swamping is unlikely in the Coast Range, where serpentine is much more extensive and contiguous. The phenological problems facing serpentine univoltines are already impressive, without the problems of maladaptive gene flow. The constraints operating on them resemble the dilemma of another vernal-univoltine butterfly, *Pieris virginiensis* Edwards (Pieridae) (Cappuccino & Kareiva, 1985). If serpentine *zelicaon* emerge too early in spring, they run the risk of catastrophe (usually associated with cutoff "cold lows," a frequent but unpredictable feature of California spring weather) which can wipe out the year's reproductive effort. (We have seen this happen on ridgetops in Napa County subjected to subfreezing temperatures and 14-cm April snows.) If they emerge too late, however, the risk of "cold low" damage is reduced at the cost of increased risk that the host plants will senesce before the larvae are full-grown. (We've seen that, too.) The response of these populations seems to be a combination of a protracted emergence in any given year, and a high proportion (over 30%) of multiple-year diapausing pupae, with some survival and successful eclosion even after 5 yr. Both of these traits "spread the risk," but at the cost of high risk to the individual (protracted emergence) and substantially increased generation time (multiple-year diapause). The genetics underlying them is difficult to study given the combination of time lags and the uncertain relation between lab and field conditions as physiological signals. But one can predict that the introgression of genes favoring direct development would be highly disruptive to the local phenology and subject to intense negative selection. Strong selection on life-history traits across an edaphic boundary could even lead to incipient speciation (Endler, 1973).

THE NEXT CALIFORNIAN ECOTYPE?

The attractiveness of both sweet fennel and citrus lies mostly in their availability through the long, hot, dry summer when potential hosts are rare and/or spottily distributed. Two species of the umbelliferous genus *Ammi* (*A. majus* L. and *A. Visnaga* [L.] Lam.) are currently spreading in California; both are summer annuals which could extend host availability into August where they occur. Like the naturalized winter annual poison hemlock (*Conium*), they could not in themselves sustain multivoltinism, but they could allow for two to three broods per year, especially in combination with a vernal host. It is thus striking that they do not seem to be used afield.

Laboratory studies (M. Nelson, S. Graves, A. M. Shapiro, unpublished) show that both *Ammi* are highly toxic to multivoltine Sacramento Valley *zelicaon* larvae in culture. Field trials on *A. Visnaga* in a site where it grows intermixed with both fennel and poison hemlock show extremely poor survival in comparison with larvae on the others. In this site *Ammi* has been present less than 5 yr, and at least some females still lay on it. In fact, oviposition trials in the lab indicate that most females oviposit on *Ammi* quite readily.

This situation has at least two potential outcomes: either *Ammi* toxicity will keep it outside the dietary of *P. zelicaon*, or (most likely where other summer hosts are unavailable, by analogy to the origin of the citrus ecotypes) selection for *Ammi* tolerance will lead to a bi- or trivoltine or host-switching multivoltine ecotype where none had been before.

Probably both positive and negative selection responses to *Ammi Visnaga* are happening right now. Meanwhile, *Ammi majus* is following I-5 from Stockton north toward Sacramento and has leapfrogged since 1990 to an I-80 interchange at Vacaville. *A. Visnaga* continues a slow creep over the southern Sacramento Valley apparently aided by the transport of caked mud on earth-moving machinery, so plenty of evolutionary opportunities can be expected. There is a substantial and growing literature on the toxicology, detoxification mechanisms and metabolism of secondary compounds from Umbelliferae (Berenbaum, 1983; Ivie, 1983) with implications for the ease with which accommodation to a given host is likely to occur; some are clearly going to be more difficult than others, and some ecotypes of *P. zelicaon* may be preadapted to *Ammi* by virtue of feeding on chemically similar hosts.

CODA

California *Papilio zelicaon* have been presented — by a conjunction of geography, climate and human impact — with an unusual range of opportunities to develop local life histories that match locally available host plants. The capacity to do this seems inherent in its lineage — a careful examination of Old World *machaon* will undoubtedly show many close parallels to our phenomena; *machaon* populations occur in Mediterranean and various arid and semiarid climates. But the evolution of new ecotypes is a local process which depends on the integration of genetic and physiological potentials with ecological conditions (the dispersion and seasonal availability of hosts) and the ability to handle hosts metabolically. This is an ongoing process which has been encouraged by the development of new plant communities since the Pleistocene, the introduction and spread of new potential hosts from elsewhere, and agricultural practices. Although not reflected at a taxonomic level, the situation is fluid: ecotypic diversity has not yet reached its full potential in California.

ACKNOWLEDGEMENTS

Research on *P. zelicaon* has been supported by California Agricultural Experiment Station Projects CA-D*-AZO-3593 and CA-D*-AZO-4905-H. Most of the results reported here are due to the work of Steve Sims, Ken Masuda, Mark Tong, Marj Nelson and Sherri Graves, all of whom I thank effusively. John Thompson, May Berenbaum and Paul Feeny have been very forthcoming over the years, which is also greatly appreciated. Joel Kingsolver and Bob Pyle made helpful editorial suggestions, not all of which I have followed.

Literature Cited

Bakker, E. 1971. *An Island Called California.* Univ. California Press, Berkeley and Los Angeles. 357 pp.

Berenbaum, M. 1983. Coumarins and caterpillars: A case for coevolution. *Evolution* 37:163-179.

Boisduval, J. B. A. 1868. Lepidoptères de la Californie. *Ann. Soc. Ent. Belgique* 12:5-94.

Bowden, S. R. 1953. Timing of imaginal development in male and female hybrid Pieridae (Lep.). *Entomologist* 86:255-264.

Bowden, S. R. 1957. Diapause in female hybrids: *Pieris napi adalwinda* and related subspecies (Lep.). *Entomologist* 90:247-254, 273-281.

Bradshaw, A. D. 1971. Plant evolution in extreme environments. Pages 20-50 *in* R. Creed, ed., *Ecological Genetics and Evolution.* Blackwell, London.

Cappuccino, N. and P. Kareiva. 1985. Coping with a capricious environment: A population study of a rare Pierid butterfly. *Ecology* 66:152-161.

Clarke, C. A. and P. M. Sheppard. 1970. Is *Papilio gothica* a good species? *Jour. Lepid. Soc.* 24:230-233.

Coolidge, K. R. 1990. A California orange dog. *Pomona Coll. Jour. Entomol.* 2:333-334.

Dennis, R. L. H. 1977. *The British Butterflies: Their Origin and Establishment.* E.W. Classey, Oxford. 318 pp.

Dethier, V. G. 1941. Chemical factors determining the choice of food plants by *Papilio* larvae. *American Nat.* 75:61-73.

Dobzhansky, Th. 1970. *Genetics of the Evolutionary Process.* Columbia Univ. Press, New York. 505 pp.

Emmel, J. F. and O. Shields. 1979. Larval foodplant records for *Papilio zelicaon* in the western United States and further evidence for the conspecificity of *P. zelicaon* and *P. gothica*. *Jour. Res. Lepid.* 17:56-67.

Endler, J. 1973. Gene flow and population differentiation. *Science* 179:243-250.

Frenkel, R. E. 1970. Ruderal vegetation along some California roadsides. *Univ. California Publ. Geogr.* 20:1-163.

Hittell, J. S. 1866. *The Resources of California*. A. Roman, San Francisco, 494 pp.

Hopkins, A. D. 1917. Entomologists' discussion. *Jour. Econ. Entomol.* 10:92-93.

Horton, J. R. 1922. A swallowtail butterfly injurious to California orange trees. *Monthly Bull. U.S. Bur. Entomol.* 11:377-387.

Ivie, G. W., D. L. Bull, R. C. Beier, N. W. Pryor, and H. Oertli. 1983. Metabolic detoxification: Mechanisms of insect resistance to plant psoralens. *Science* 221:374-376.

Little, T. M. and F. J. Hills. 1975. *Statistical Methods in Agricultural Research*. Univ. California, Davis. 242 pp.

Masuda, K. K. 1981. The effects of two host plants, *Foeniculum vulgare* Mill. and *Citrus sinensis* Osbeck, on adult oviposition behavior and larval and pupal biology of the anise swallowtail butterfly (*Papilio zelicaon* Lucas) (Lepidoptera: Papilionidae). *M.S. Thesis, Zoology*. Univ. California, Davis. 82 pp.

Munz, P. A. and D. D. Keck. 1970. *A California Flora*. Univ. California Press, Berkeley. 1681 pp.

Oliver, C. G. 1972. Genetic and phenotypic differentiation and geographic distance in four species of Lepidoptera. *Evolution* 26:221-241.

Opitz, K. W. and R. G. Platt. 1969. *Citrus Growing in California*. California Agri. Expt. Sta., Ext. Serv. 59 pp.

Remington, C. L. 1968. A new sibling *Papilio* from the Rocky Mountains, with genetic and biological notes. *Postilla* 119:1-40.

Robbins, W. W. 1940. Alien Plants Growing Without Cultivation in California. *Univ. California Agri. Expt. Sta., Bull.* 637. 128pp.

Sculley, R. 1973. The natural state. *Davis New Rev.* 1:3-13.

Shapiro, A. M. 1974. Ecotypic variation in montane butterflies. *Wasmann Jour. Biol.* 32:267-280.

Shapiro, A. M. 1975. *Papilio "gothica"* and the phenotypic plasticity of *P. zelicaon* (Papilionidae). *Jour. Lepid. Soc.* 29:79-84.

Shapiro, A. M. 1980. Mediterranean climate and butterfly migration: An overview of the California fauna. *Atalanta* 11:181-188.

Shapiro, A. M. 1984. Non-diapause overwintering by *Pieris rapae* (Lepidoptera: Pieridae) and *Papilio zelicaon* (Lepidoptera: Papilionidae) in California: Adaptiveness of Type III diapause-induction curves. *Psyche* 91:161-169.

Shields, O. 1967. Hilltopping: An ecological study of summit congregation behavior of butterflies on a southern California hill. *Jour. Res. Lepid.* 6:69-78.

Sims, S. R. 1980. Diapause dynamics and host plant suitability of *Papilio zelicaon* (Lepidoptera: Papilionidae). *American Midl. Nat.* 103:375-384.

Sperling, F. 1987. Evolution of the *Papilio machaon* group in western Canada (Lepidoptera: Papilionidae). *Quaest. Entomol.* 23:198-315.

Thompson, J. 1986. Patterns in coevolution. Pages 119-143. *in* A. R. Stone & D. J. Hawksworth, eds., *Coevolution and Systematics.* Oxford Univ. Press, Oxford.

Thompson, J. 1988a. Variation in preference and specificity in monophagous and oligophagous swallowtail butterflies. *Evolution* 42:118-128.

Thompson, J. 1988b. Evolutionary genetics of oviposition preference in swallowtail butterflies. *Evolution* 42:1223-1234.

Thompson, J. 1988c. Evolutionary ecology of the relationship between oviposition preference and performance of offspring in phytophagous insects. *Entomol. Exp. Appl.* 47:3-14.

Thompson, K. 1961. Riparian forests of the Sacramento Valley, California. *Ann. Assoc. American Geogr.* 51:294-314.

Tong, M. L. and A. M. Shapiro. 1989. Genetic differentiation among California populations of the anise swallowtail butterfly, *Papilio zelicaon* Lucas (Papiliondae). *Jour. Lepid. Soc.* 43:217-228.

Wiklund, C. 1974. The concept of oligophagy and the natural habitat and host plants of *Papilio machaon* L. in Fennoscandia. *Entomol. Scand.* 5:151-160.

Wiklund, C. 1981. Generalist vs. specialist oviposition behavior in *Papilio machaon* (Lepidoptera) and functional aspects of the hierarchy of oviposition preferences. *Oikos* 36:163-170.

Multi-character Ecotypic Variation in Edith's Checkerspot Butterfly

Michael C. Singer[1], Raymond R. White[2], Daniel A. Vasco[1], Christian D. Thomas[3], and David A. Boughton[1]

[1] Department of Zoology, University of Texas, Austin, TX 78712; [2] H. P. Harvey & Associates, P.O. Box 1180, Alviso, CA 95002; [3] School of Biological Sciences, University of Birmingham, Edgbaston, Birmingham, U.K.

Ecotype formation should be best developed in species that live in small, stable habitat patches with low gene flow among them. Conversely, when movement among patches is frequent and patch size small, we expect selection for local adaptation to be swamped by gene flow. When selection pressures within a patch vary through time, ecotype formation may also be prevented. For this model of ecotype formation, butterflies provide a useful set of examples because they vary greatly in the extent of gene flow, and because of the sorts of spatial and temporal variation in selection pressures that they experience. At one extreme, the highly dispersive monarch butterfly develops regional genetic variation each summer, only to have it wiped out each fall by migration and coalescence of populations from different areas. In contrast, our study organism, Edith's checkerspot, typically occurs as small, local populations which show discernible ecotypic variation over distances of 200 m and dramatic variation over distances on the order of 20 km. We document extensive interpopulation variation in two categories of traits: (1) life history traits such as egg size, egg cluster size and adult size and (2) adaptations to host plants, such as habitat choice and host preference. In most, but not all, cases, the patterns of variation in these traits are clearly associated with environmental variables.

Although the concept of ecotypic variation has its roots in botany (Turesson, 1922; Clausen, Keck & Hiesey, 1940), interpopulation genetic variation within animal species resembles that in plants and deserves equivalent treatment. Animals show both large-scale patterns or clines (Endler, 1977) correlated with altitude or latitude (Pickens, 1965) and small-scale local adaptation to soil type or habitat structure that produces patchworks of phenotypes. As an example of an altitudinal trend, our own study organism, the checkerspot butterfly *Euphydryas editha*, decreased in size with increasing altitude (see below) along the same transect through the Sierra Nevada Mountains that was used by Clausen, Keck & Hiesey (1958) to show the same phenomenon in *Achillea millefolium*. As an example of a patchwork, the lizard, *Anolis oculatus*, showed local patterns of variation in body size, shape, color pattern and scalation that were correlated with local environment and maintained by habitat-specific natural selection (Malhotra & Thorpe, 1991). This selection generated at least four ecotypes in the island of Dominica, with ecotypic variation occurring over very small distances, presumably in spite of gene flow between lizards in different habitat types. Likewise, guppies showed ecotypic variation of color pattern between habitats differing in size of substrate particles (Endler, 1980). Animals may even show ecotypic variation in communication systems; for example, the advertisement calls of cricket frogs (*Acris crepitans*) in Texas vary among habitat types. When these calls were played in forest habitats, those typical of frogs from forest habitats traveled with less degradation than calls typical of frogs from open habitats (Ryan, *et al.*, 1990). These animal examples parallel work on plants in which local genetic adaptation to soil type or microclimate creates a spatial patchwork (Warwick & Briggs, 1978; Davy, 1988; Kruckeberg, this volume). Both plants and animals also show local adaptation to the presence or activity of other organisms. For example, populations of the grass *Poa annua* vary in height in consequence of selection by grazing herbivores (Warwick & Briggs, 1978) and local variation in color pattern of *Heliconius* butterflies results from selection pressures associated with the distribution of co-mimetic species rather than from climatic or edaphic factors (Mallet & Barton, 1989).

Spatially heterogeneous patterns of selection should create the potential for ecotype formation. The primary forces countering this potential should be genetic drift when populations are small, and gene flow when migration and dispersal are frequent (Fisher, 1950; Ehrlich & Raven, 1969; Slatkin, 1973). Thus, when gene flow is high and habitat patch size small, we may expect tendencies for local adaptation to be swamped. Conversely, sedentary organisms inhabiting large habitat patches have more opportunity to evolve local adaptation. This simple model becomes complex, however, when one considers that selection pressures on any particular population can vary in direction through time. When selection is inconsistent through time, it may prevent ecotypic formation as effectively as does gene flow from populations experiencing different selection pressures.

The potential for ecotype formation, then, depends critically on spatial and temporal patterns of selection among populations.

Like ecotype formation, the rate of gene flow is also subject to selection, via selection for dispersal. In one class of models, the evolution of dispersal depends critically on spatial and temporal patterns of fitness among spatially separated populations, although in others the cost of dispersal, or the cost of inbreeding and outbreeding are more important (Johnson & Gaines, 1990). A set of generalizations arises from the former class of models: First, spatial variation without temporal variation reduces dispersal rate, because there is no advantage to dispersing, and local adaptation may mean there is a cost to dispersing. Second, temporal variation in patches that is uncorrelated among patches tends to favor dispersal, because when the currently occupied patch is poor, dispersal may result in occupying a better patch (Johnson & Gaines, 1990). Thus, dispersal may evolve in response to the same sorts of selection pressures as ecotypes do (*i.e.*, spatial and temporal variation in fitness), but is favored when ecotype formation is disfavored: when spatial variation is low and temporal variation high.

Ecotype formation and increases in dispersal rate may be opposed to each other. For example, in Gill's (1978) model, increased competitive ability and dispersal are alternative strategies whose relative advantage depends on spatial and temporal variation in crowding. So, in this model, crowding results in either greater dispersal or locally increased competitive ability. The result depends on the nature of other patches that are available. Whichever result occurs tends to be self-reinforcing: because dispersal is frequently the conduit of gene flow, situations favoring it may tend to preclude the formation of ecotypes. Likewise, local adaptation will itself tend to generate selection reducing emigration and hence gene flow, thereby enhancing local differentiation (Gilbert & Singer, 1973). In this respect, butterflies provide a useful set of examples because they vary greatly in the extent of population subdivision. For example, the monarch butterfly (*Danaus plexippus*) develops regional genetic variation each summer, only to have it wiped out each fall by migration and coalescence of populations from different areas (Eanes & Koehn, 1979). The less dispersive tiger swallowtail, *Papilio glaucus*, shows patterns of adaptation to latitudinal variation of growing season length on a scale of hundreds of kilometers (Scriber & Lederhouse,

1992). Our study organism, Edith's checkerspot (*Euphydryas editha*), is more sedentary, and probably more representative of most butterflies (Ehrlich, 1961; Gilbert & Singer, 1973; Harrison, *et al.*, 1988). It shows discernible ecotypic variation over distances of 200 m and dramatic variation over distances on the order of 20 km. Below, we describe this variation in two categories of traits: (1) life history traits such as egg size, egg cluster size and adult size and (2) adaptations to host plants, such as habitat choice, and host preference. In spite of the extent of host-associated differentiation among populations, *Euphydryas* show no tendency to speciate following acquisition of a novel host. In this sense they behave very differently from other well-studied herbivorous insects, such as the true fruit flies, *Rhagoletis* (Bush, 1969). In comparison with *Euphydryas, Rhagoletis* show a greater tendency to speciate and a reduced tendency for intraspecific ecotype formation.

STUDY INSECT

E. editha fits the model of Fox & Morrow (1981), who gathered examples to show that conspecific insect populations frequently have very different diets, leading to the generation of much greater diet breadth at the species than at the population level. *E. editha* occurs in scattered, semi-isolated populations (Ehrlich, *et al.*, 1975; Baughman, *et al.*, 1990) in which adult insects oviposit on plants in the families Plantaginaceae and/or Scrophulariaceae. Some populations of this insect are monophagous, while others use up to four host genera. As far as is known, existing *E. editha* populations are univoltine, with a single generation each year. Adults fly in March-April at sea level, April-May at 6,000 feet (1800 m) elevation, May-June at 8,000 feet (2400 m), and July-August at 12,000 feet (3600 m).

Eggs are laid in clusters during the flight season and hatch after 1-2 weeks. Young larvae feed gregariously on the plants that their mothers chose, becoming more independent and capable of searching for hosts when they enter third instar at about ten days old. Because these young larvae are tied to the hosts chosen by their mothers, their food choice behavior is not important and oviposition choice decisions have large effects on fitness (Singer 1984; Moore 1989).

If second instar or very early third instar larvae cannot find food, they starve, but after a few days' feeding in third instar, they reach a stage at which they respond to lack of food by diapause. If food remains plentiful, larvae enter an obligatory nine-month diapause at the end of third or fourth instar. After diapause is broken either by winter rains (at sea level), or by snowmelt (at high elevation), larvae feed for a few weeks before wandering to find pupation sites under litter or inside pine cones. Mating is not associated with host plants, and usually occurs at the site of female emergence (Singer & Thomas, 1992). Oviposition usually begins on the second or third day of the female's adult life. In many of our study populations, each female lays one egg cluster per day, but in populations where cluster size is small, oviposition is more frequent.

The host plants of *E. editha* are all herbaceous. *Pedicularis* spp., *Castilleja* spp., *Penstemon* and *Plantago lanceolata* are perennial, while *Collinsia* are small, erect annuals. Chemically, these four plant genera have at least one common trait in that they all contain iridoid glycosides, but experiments in which insects were offered plant extracts have yet to provide evidence that iridoids are used by these butterflies to identify hosts (Singer, *et al.*, unpublished). *Pedicularis* and *Castilleja* are hemiparasitic and obtain secondary chemicals from their hosts. However, *E. editha* failed to discriminate among *Pedicularis* plants that had acquired alkaloids from lupines and those that had not (Stermitz, *et al.*, 1989). So, as yet we have only negative evidence about the chemical traits used by *E. editha* to identify hosts.

WHAT TRAITS SHOW ECOTYPIC VARIATION?

Life History Traits: Adult Size, Egg Size, Egg Cluster Size

We have described *E. editha* as an insect with but a single generation per year. There are reports of populations with facultative second generations in San Diego (F. Thorne, pers. commun.) and around Mono Lake (S. O. Mattoon, pers. commun.), but these populations are apparently now extinct. When *Papilio zelicaon* is presented with a long season in which food for larvae is available, it responds by changes in diapause that increase the number of generations per year (Shapiro, this volume). Although *E. editha* does not now show ecotypic variation in physiological traits that influence the number of generations per year, it does vary in size, which may be a different response to precisely the same environmental variable (available time for larval growth) that affects *P. zelicaon*. We suspect that, when sufficient time is available, *E. editha* increases its body size, and, hence, its fecundity. Habitats at low elevation usually have longer growing seasons than high elevation sites, and should support faster larval growth because of warmer night-time temperatures. These effects could be responsible for the observed negative correlation between insect size and altitude (Fig. 1). These differences among populations in size are replicated in laboratory cultures in which all insects are raised on the same host species, so we can conclude that they are genetically based. We have no demonstration that selection pressures associated with duration of host availability are responsible for the correlation shown in Fig. 1; this is merely a plausible explanation.

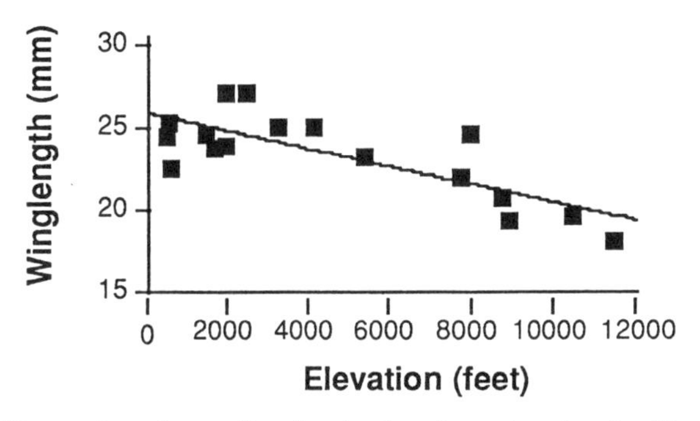

FIGURE 1. Regression of mean female wing length on elevation for 17 populations of *E. editha* in California and Nevada. Elevation explains 62% of the variation in mean wing length.

Egg cluster size is also genetically variable among populations, and this variation has an obvious environmental correlate: host affiliation. Insects that feed on *Plantago erecta* or *Pedicularis* spp. lay large clusters averaging 45-90 eggs. Those that feed on *Collinsia tinctoria* lay smaller clusters averaging about 25 eggs, while on *Collinsia torreyi* or *C. greenei*, the average is only 4-6 eggs per cluster. Experiments (unpublished) in which egg cluster size has been manipulated in the field have shown a significant interaction between group size, larval survival and host species that could be the basis for the generation of host-associated ecotypic variation of egg cluster size.

Egg size variation among populations is apparently independent of variation in adult size and not strongly correlated with host affiliation. The only environmental factor that we have detected that is correlated with this variation in egg size is latitude: there is a weak tendency for insects in southern populations to lay small eggs.

HOST PREFERENCE TRAITS: VISUAL, CHEMICAL AND PHYSICAL RESPONSES.

Before discussing traits that show host-associated ecotypic variation, we need to describe the oviposition behavior of our study insect. Female *E. editha* that are searching for oviposition sites alight on all plant species present, both hosts and non-hosts. Parmesan *et al.* (MS) recorded these alights and calculated an alighting bias for each plant species as the ratio of observed to expected alight frequencies, with the expected frequency calculated from the proportional representation of the plant in the vegetation (random transects). Alights were not random; alighting biases were often strong, and varied among plant species. Over 80% of this variation in alighting bias was explained by the physical traits of the plant: in the population studied, insects were attracted to plants with complex leaves and rosette growth form, whether or not they were hosts. This evidence indicates that searching *E. editha* are attracted to plants on the basis of visual stimuli. The nature of these responses to visual stimuli varies among populations (Parmesan, 1991), but we have no evidence as yet on the mechanisms that generate this variation.

After alighting, female *E. editha* taste the plant surface by drumming it with their atrophied foretarsi. A positive response to this tasting constitutes curling of the abdomen and extrusion of the ovipositor, which is then pressed hard against the lower surface of the leaf or flower. We have shown that this abdominal curling is a response to plant chemicals. Differences among butterfly populations in responses to intact plants are maintained in responses to extracts of these same plants on filter paper (M. C. Singer & M. Perez, unpublished). If a plant is chemically acceptable, the final stage in the sequence is response to physical stimuli detected by the ovipositor. We deduce that the ovipositor is not sensitive to chemical stimuli, since we observe that only the tarsi, not the ovipositor, must be in contact with the host for oviposition to occur. In consequence, eggs are sometimes laid adjacent to hosts rather than on them.

SCALE OF ECOTYPIC VARIATION IN HOST PREFERENCE

Host choice has strong effects on offspring fitness in *Euphydryas* (Rausher, 1982; Singer, 1984; Moore, 1989), but the most suitable hosts are not the same in different habitats. Where habitats are patchy, selection may favor the use of different host species in the different patch types (Singer, 1983). This process will tend to cause differentiation among local populations in host preference. Such differentiation has been detected between populations only 100-200 m apart, in spite of substantial gene flow (Singer, *et al.*, 1992). However, the identities of preference phenotypes are not different between adjacent populations, it is merely their frequencies that vary (Fig. 2). It is only when we compare over greater distances that we can find populations that have no phenotypes in common. For example, every one of more than 500 *E. editha* from Generals' Highway would accept *Pedicularis semibarbata* for oviposition, while none of 55 from Kings Canyon would do so (Vasco *et al.*, unpublished). These two sites are about 13 km apart. For these particular examples, our evidence that interpopulation variation of preference is genetic stems from duplication of this variation in insects raised in captivity on the same host species. However, we have performed reciprocal crosses between another pair of populations with different preferences, and shown that the differences were genetically based (Singer, *et al.*, 1991).

Granted that extensive interpopulation variation of host affiliation exists in *E. editha* (White & Singer, 1973), and that gene flow among populations is restricted, we expect to find interpopulation variation of physiological performance of larvae on different hosts. This ex-

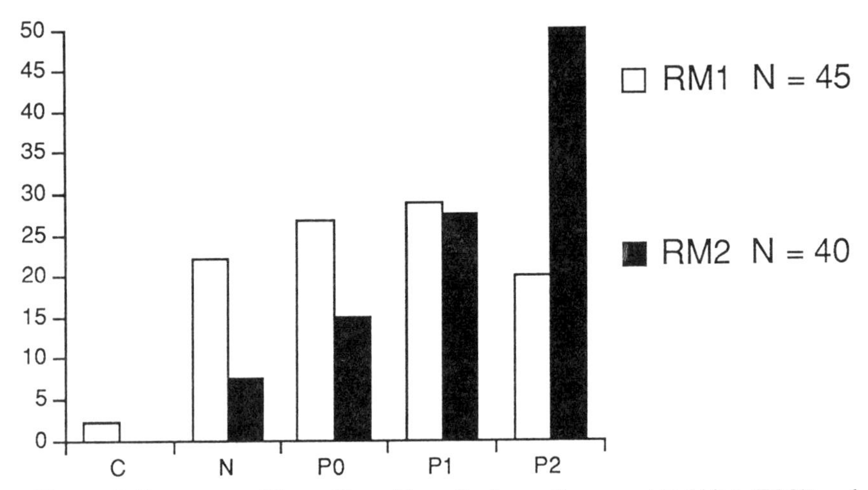

FIGURE 2. Percentage of butterflies with particular preferences at Rabbit 1 (RM1) and Rabbit 2 (RM2) in 1992.

C = preference for *Collinsia torreyi* over *Pedicularis semibarbata*; N = no preference
PO = preferring *P. semibarbata*, but accepting *C. torreyi* on the same day (day 1)
P1 = preferring *P. semibarbata*, and not accepting *C. torreyi* until day 2
P2 = preferring *P. semibarbata*, and still rejecting *C. torreyi* on day 2

pectation is fulfilled (Rausher, 1982). Other less obvious traits may also be genetically variable among populations using different host species. These include preference for flying in shade vs. sun, which brings adults into contact with different plant species (Gilbert & Singer, 1973), and larval geotropism, which brings larvae into contact with edible parts of the plants on which they hatch from the eggs (Singer, in press). For example, in populations feeding on *Pedicularis* spp. adults drop to the base of the plant to lay eggs, and larvae tend to be positively geotropic, spinning webs between the ground and the lower leaves of the rosette. In contrast, *Collinsia torreyi* does not possess edible parts at ground level. Insects ovipositing on *C. torreyi* place their eggs close to the top of the plant and produce larvae that remain close to the meristems, feeding on buds, flowers and developing seeds.

INTERACTION BETWEEN HOST PREFERENCE AND LIFE HISTORY TRAITS

The swallowtail *Papilio zelicaon* responds to increases in available time (length of season) by evolution of diapause traits, thereby increasing the number of generations per year (Shapiro, this volume). Its relative, *Papilio glaucus* responds either by increasing the number of generations per year and/or by genetically-based increase in diet breadth (Scriber & Lederhouse, 1992). When possible, the diet expands to include species on which development is slow; when time is severely constrained, the diet contracts to the set of host species on which development is fastest. Thus, there is a potential interaction between the evolution of diet and of life-history traits. If we apply this reasoning to *E. editha*, we expect that in habitats where hosts are available to larvae for long periods, the insects should have either broad diet and/or several generations per year. In fact, *E. editha* do not vary in the number of generations per year, and their variation in diet breadth is not obviously related to the length of time available for larval growth. Instead, these insects vary in size, as we have described, being smallest in the high elevation habitats where summers are very short. This example shows us that the Scriber & Lederhouse (1992) model is an incomplete description of the options available to the insects. Whatever the generality of this model, it foreshadows ways in which studies of relationships between ecotypic variation of diet and of life-history traits will help us to understand the evolution of both sets of characters.

Autotrophic plants do not have host preferences, but they do have physiological adaptations to particular substrates. Like animals, they may respond to spatial patchiness with either plastic or genetic variation in life-history (Law, *et al.*, 1977) and/or substrate adaptation (Jain & Bradshaw, 1966; Antonovics, *et al.*, 1971; Bradshaw 1977). Thus, they provide the raw material for development of theories that would predict relationships between ecotypic variation of life-history and substrate adaptation, as in the example of *Papilio glaucus* described above.

INTERACTION BETWEEN PLANT AND INSECT VARIATION

The extensive variation of oviposition preference among populations of *E. editha* is not the only cause of geographical differences in diet. A plant species may be excluded from the diet simply because it is locally absent. This can be demonstrated by taking the plant to the habitat and showing that the local *E. editha* would oviposit on it, if it grew there naturally (Thomas, *et al.*, 1987). A second mechanism that excludes plants from the diet is resistance to attack in the local plant population. Just as preference varies among insect populations, acceptability to the ovipositing insects varies among plant populations (Singer & Parmesan, 1993). At one study site, Sonora Junction (Mono Co., California), *Penstemon rydbergii* was excluded from the diet of *E. editha* by a combination of local preference of the insects and local acceptability of the plant. At another site, Frenchman Lake (Plumas Co., California), *P. rydbergii* was the principal host. Here, the insects were genetically more *Penstemon*-preferring and the *P. rydbergii* were genetically more acceptable than those at Sonora Junction. So, the observed pattern of association among plants and insects is an emergent property arising from interaction between variation among plant populations in resistance and variation among insect populations in preference (Singer & Parmesan, 1993). The study of ways in which ecotypic variation influences species interactions is still in its infancy, but this example shows that it has the potential to elucidate the mechanisms by which the interactions themselves may vary among sites.

References

Antonovics, J., A. D. Bradshaw, and R. G. Turner. 1971. Heavy metal tolerance in plants. Pages 1-85 *in* J. B. Cragg, ed., *Advances in Ecological Research.* Academic Press, Inc., New York.

Bradshaw, A. D. 1977. Some of the evolutionary consequences of being a plant. *Evol. Biol.* 5:25-47.

Baughman, J. F., P. F. Brussard, P. R. Ehrlich, and D. D. Murphy. 1990. History, selection, drift and gene flow: Complex differentiation in checkerspot butterflies. *Canadian Jour. Zool.* 68:1967-1975.

Bush, G. L. 1969. Sympatric host race formation and speciation in frugivorous flies of the genus *Rhagoletis. Evolution* 23:237-251.

Clausen, J., D. D. Keck, and W. M. Hiesey. 1940. Experimental Studies on the Nature of Species. l. Effect of Varied Environments on Western North American Plants. *Carnegie Inst. Washington, Publ.* 520.

Clausen, J., D. D. Keck and W. M. Hiesey. 1958. Experimental Studies on the Nature of Species. III. Environmental Responses of Climatic Races of *Achillea. Carnegie Inst. Washington, Publ.* 581.

Davy, A. J. 1988. Plant Population Ecology. *Symp. British. Ecol. Soc.* 28:1-23.

Eanes, W. F. and R. K. Koehn. 1978. An analysis of genetic structure in the Monarch butterfly, *Danaus plexippus. Evolution* 32:787-797.

Ehrlich, P. R. 1961. Intrinsic barriers to dispersal in checkerspot butterfly. *Science* 134:108-109.

Ehrlich, P. R. and P. H. Raven. 1969. Differentiation of populations. *Science* 165:1228-1231.

Ehrlich, P. R., R. R. White, M. C. Singer, S. W. McKechnie and W. B. Watt. 1975. Checkerspot butterflies: A historical perspective. *Science* 176:221-228.

Endler, J. A. 1977. *Geographic Variation, Speciation and Clines.* Princeton Univ. Press, Princeton, New Jersey.

Endler, J. A. 1980. Natural selection on color patterns in *Poecilia reticulata. Evolution* 34:76-91.

Fisher, R. A. 1950. Gene frequencies in a cline determined by selection and diffusion. *Biometrics* 6:353-361.

Fox L. R. and P. A. Morrow. 1981. Specialization:species property or local phenomenon? *Science* 211:887-893.

Gilbert, L. E. and M. C. Singer. 1973. Dispersal and gene flow in a butterfly species. *American Nat.* 107:58-72.

Gill, D. E. 1978. On selection at high population density. *Ecology* 59:1289-1291.

Harrison, S., D. D. Murphy and P. R. Ehrlich. 1988. Distribution of the Bay Checkerspot Butterfly *Eupydryas editha bayensis*: Evidence for a metapopulation model. *American Nat.* 132:360-382.

Jain, S. J. and A. D. Bradshaw. 1966. Evolutionary divergence among adjacent plant populations. I. The evidence and its theoretical analysis. *Heredity* 21:407-441.

Johnson, M. L. and M. S. Gaines. 1990. Evolution of dispersal: Theoretical models and empirical tests using birds and mammals. *Ann. Rev. Ecol. Syst.* 21:449-480.

Law, R., A. D. Bradshaw and P. D. Putwain. 1977. Life-history variation in *Poa annua*. *Evolution* 31:233-246.

Malhotra, A. S. and R. S. Thorpe. 1991. Experimental detection of rapid evolutionary response in natural lizard populations. *Nature* 353:347-348.

Mallet, J and N. H. Barton. 1989. Strong natural selection in a warning-color hybrid zone. *Evolution* 43:421-431.

Moore, W. S. 1977. An evaluation of narrow hybrid zones in vertebrates. *Quart. Rev. Biol.* 52:253-278.

Moore, S. D. 1989. Patterns of juvenile mortality within an oligophagous butterfly population. *Ecology* 70:1726-1731.

Parmesan, C. 1991. Evidence against plant “apparency” as a constraint on evolution of insect search efficiency. *Jour. Insect Behav.* 4:417-430.

Pickens, P. E. 1965. Heart rate of mussels as a function of latitude, intertidal height, and acclimation temperatures *Physiol. Zool.* 38:390-405.

Rausher, M. D. 1982. Population differentiation in *Euphydryas editha* butterflies: Larval adaptation to different hosts. *Evolution* 36:581-590.

Ryan, M. J., R. B. Cocroft and W. Wilczynski. 1990. The role of environmental selection in intraspecific divergence of mate recognition signals in the cricket frog, *Acris crepitans*. *Evolution* 44:1869-1872.

Scriber, J. M. and R. C. Lederhouse. 1992. The thermal environment as a resource dictating patterns of feeding specialization of insect herbivores. Pages 429-465 *in* M. D. Hunter, T. Ohgushi, & P. W. Price, eds., *Effects of Resource Distribution on Animal-Plant Interactions*. Academic Press, New York & London.

Shapiro, A. M. 1995. From the mountains to the prairies to the ocean white with foam, *Papilio zelicaon* makes itself at home. This volume.

Singer, M. C. 1983. Determinants of multiple host use by a phytophagous insect population. *Evolution* 37:389-403.

Singer, M. C. 1984. Butterfly-hostplant relationships: Host quality, adult choice and larval success. Pages 81-88 *in* R. Vane-Wright & P. Ackery, eds., *The Biology of Butterflies*. Roy. Entomol. Soc. London, Symp. No. 13.

Singer, M. C. (in press). Behavioral constraints to the evolution of insect diet breadth. Pages 279-296 *in* L. Real, ed., *Behavioral Mechanisms in Evolutionary Ecology*. Univ. Chicago Press.

Singer, M. C., R. A. Moore and D. Ng. 1991. Genetic variation in oviposition preference between butterfly populations. *Jour. Insect Behav.* 4:531-535.

Singer, M. C., D. Ng, D. Vasco and C. D. Thomas. 1992. Rapidly evolving associations among oviposition preferences fail to constrain evolution of insect diet. *American Nat.* 139:9-20.

Singer, M. C. and C. Parmesan. 1993. Sources of variation in patterns of plant-insect association. *Nature* 361:251-253.

Singer, M. C. and C. D. Thomas 1992. The difficulty of deducing behavior from resource use: An example from hilltopping in checkerspot butterflies. *American Nat.* 140:654-664.

Singer, M. C., D. Vasco, C. Parmesan, C. D. Thomas and D. Ng. 1992. Distinguishing between preference and motivation in food choice: An example from insect oviposition. *Animal Behav.* 44:463-471.

Slatkin, M. 1973. Gene flow and selection in a cline. *Genetics* 75:733-756.

Stermitz, F. R., G. L. Belovsky, D. Ng and M. C. Singer. 1989. Quinolizidine alkaloids acquired by *Pedicularis semibarbata* (Scrophulariaceae) from *Lupinus perennis* (Leguminosae) fail to influence the specialist herbivore *Euphydryas editha*. *Jour. Chem. Ecol.* 15:2521-2529.

Thomas, C. D., D. Ng, M. C. Singer, J. L. B. Mallet, C. Parmesan and H. L. Billington. 1987. Incorporation of a European weed into the diet of a North American herbivore. *Evolution* 41:892-901.

Turesson G. 1922. The genotypical response of the plant species to the habitat. *Hereditas* 3:211-250.

Warwick, S. I. and D. Briggs. 1978. The genecology of lawn weeds. 1. Population differentiation in *Poa annua* in a mosaic environment of bowling green lawns and flower beds. *New Phytol.* 81:711-723.

White, R. R. and M. C. Singer.1974. Geographical distribution of hostplant choice in *Euphydryas editha*. *Jour. Lepid. Soc.* 28:103-107.

Geographical Differentiation in *Crepis tectorum* (Asteraceae): Past and Current Patterns of Selection

Stefan Andersson
Department of Systematic Botany
University of Lund, Ö. Vallgatan 18-20, S-223 61, Lund, Sweden

Comparative and demographic studies were performed to reveal past and current selection in *Crepis tectorum* (Asteraceae), a species which exhibits extensive ecogeographical variation in Europe. The major evolutionary trend reflects changes which accompanied the colonization of rocky outcrops in the Baltic lowland areas, while the origin of a widely distributed weed form probably involved smaller changes; several traits confined to the weed group are shared by more primitive species. Current patterns of selection were only partly congruent with historical selection pressures in subsp. *pumila*, a distinct outcrop form on the Baltic island of Öland. Demographic data from a natural population verified the inferred advantage of imposing seed dormancy in a habitat subject to unpredictable droughts. Direct forces of selection as revealed by multiple regression acted in favour of plants with finely divided leaves and a densely branched stem, two distinctive traits of subsp. *pumila*. Stem height was subject to selection in the "wrong" direction, indicating that weather conditions were unusual during this study. A genetic change of the population due to selection seems possible, since a heritable component of variation was found for most of the traits studied. Measurements of selection, together with evidence that patterns of variation have been stable across taxonomic levels, illuminate the connection between large-scale differentiation and present-day patterns of selection and variation at the population level. More work is needed to bridge the gap between different methodological approaches to the study of adaptation.

Because of the wide range of testable predictions provided by hypotheses about adaptation, numerous techniques have been employed to assess the role of selection in natural populations (Darwin, 1859; Endler, 1986; Primack & Kang, 1989). For instance, intraspecific genetic variation correlated with habitat differences ("ecotypic differentiation") can be used to infer past episodes of selection (Turesson, 1922; Clausen, Keck, & Hiesey, 1940; Venable, 1984), particularly if the same pattern is seen among unrelated taxa (Endler,

1986), while demographic and experimental methods yield more reliable estimates of selection in present-day populations (Endler, 1986; Primack & Kang, 1989).

It may be difficult to demonstrate the selection pressures responsible for ecotypic differentiation if populations are at or close to equilibrium, *i.e.*, if the process of adaptation occurred some time in the past. Moreover, inferences of past selection are always based on the assumption that traits evolved in situ, although information on systematic relationships may be used to identify local adaptations in the taxon investigated (Wanntorp, 1983). Directions of historical and current selection may also diverge for a trait that evolved as a non-adaptive response to selection on a correlated trait, stressing the importance of genetic analyses of characters assumed to be adaptive (Falconer, 1981). Given these and other problems (Endler, 1986), a combination of comparative, demographic and genetic studies yields more insights into the process of adaptation than each approach alone.

A comparison of past and current selection can be used to test whether demographic forces are sufficient to account for the patterns observed at higher taxonomic levels (Sokal, 1978; Felsenstein, 1988), yet relatively few authors have examined the mode of selection on traits subject to ecotypic differentiation, at least in plants (Wolff, 1988; Galen, 1989; Scheiner, 1989; Jordan, 1991; Van Tienderen & Van der Toorn, 1991). This is surprising, since plants have a number of attributes suitable for studies of adaptation (Primack & Kang, 1989). Firstly, the presence of ecologically differentiated populations is easily demonstrated in cultivation or transplant studies (Turesson, 1922; Clausen, Keck, & Hiesey, 1940; Venable, 1984) and there are relatively few problems associated with demographic studies in most plant populations. Secondly, the mating pattern of most plants can be either inferred or manipulated, making it possible to estimate the amount of genetic variation available to selection and the extent to which different traits are genetically correlated (Falconer, 1981).

In the present paper, I describe the differentiation pattern in *Crepis tectorum*, relate some of this variation to local habitat conditions, examine whether current patterns of selection correlate with historical selection pressures, and assess the role of genetic constraints on rates and patterns of population divergence. Comparative, demographic and genetic studies were performed to address the following specific questions:

1) Which traits have been subject to local selection in previous generations, *i.e.*, which traits display ecotypic differentiation? Are differences among populations consistent with predictions regarding life-history strategies in habitats with various degrees of disturbance?

2) Are traits believed to be adaptive in subsp. *pumila*, a distinct form of the species, still favoured by selection, or is selection stronger on more obvious fitness traits? Is selection on a trait (when present) direct or indirect due to correlations with other traits?

3) Have earlier episodes of selection affected the pattern and extent of plasticity in morphology and life-history, or the extent to which different leaf characters change during plant ontogeny?

4) Has the genetic architecture of ecologically important traits constrained or facilitated adaptive differentiation in *Crepis tectorum*? To what extent have distinctive traits of subsp. *pumila* evolved as a by-product of adaptive changes in genetically correlated characters?

LARGE-SCALE PATTERNS OF DIFFERENTIATION

Crepis tectorum L. (Asteraceae) is a highly polymorphic and widespread diploid (2n = 8) annual or biennial plant native to Eurasia and introduced to North America (Babcock, 1947; Najda, *et al.*, 1982). European populations have diverged for a large number of vegetative and reproductive traits, including breeding system, patterns and amounts of phenotypic plasticity and patterns of change in leaf shape during plant ontogeny (Andersson, 1989a-d, 1990a-b, 1991a-b, 1993a). The variation is geographically structured, with at least six races in Europe: a widespread weed taxon, including a weakly differentiated form adapted to arable fields in the northern regions, a weed-like form on rocky outcrops in Fennoscandia and Estonia (mostly on small islands along the coasts), and three additional outcrop forms in the Baltic region (Figs. 1, 2), which represents a centre of diversity in western *C. tectorum* (Andersson, 1990b, 1991a, 1993a). Each of these forms has a wide but fragmented distribution due to their island-like habitats (arable fields, islets, outcrops). Most of the differences seem to be of adaptive nature (see below), although patterns of genetic variation in leaf shape indicate that founder effects contribute to the diversity at the local level (Andersson, 1991a).

FIGURE 1. Drawings of plants of *C. tectorum* raised under uniform greenhouse conditions: a) weed plant from a ruderal site (the Netherlands), b) weed plant from an arable field (SE Sweden), c) outcrop plant in the mainland group (E. Sweden), d) outcrop plant in the S. Finland category, e) outcrop plant from the island of Gotland, f) outcrop plant from the island of Öland (subsp. *pumila)*. Scale bar = 10 cm.

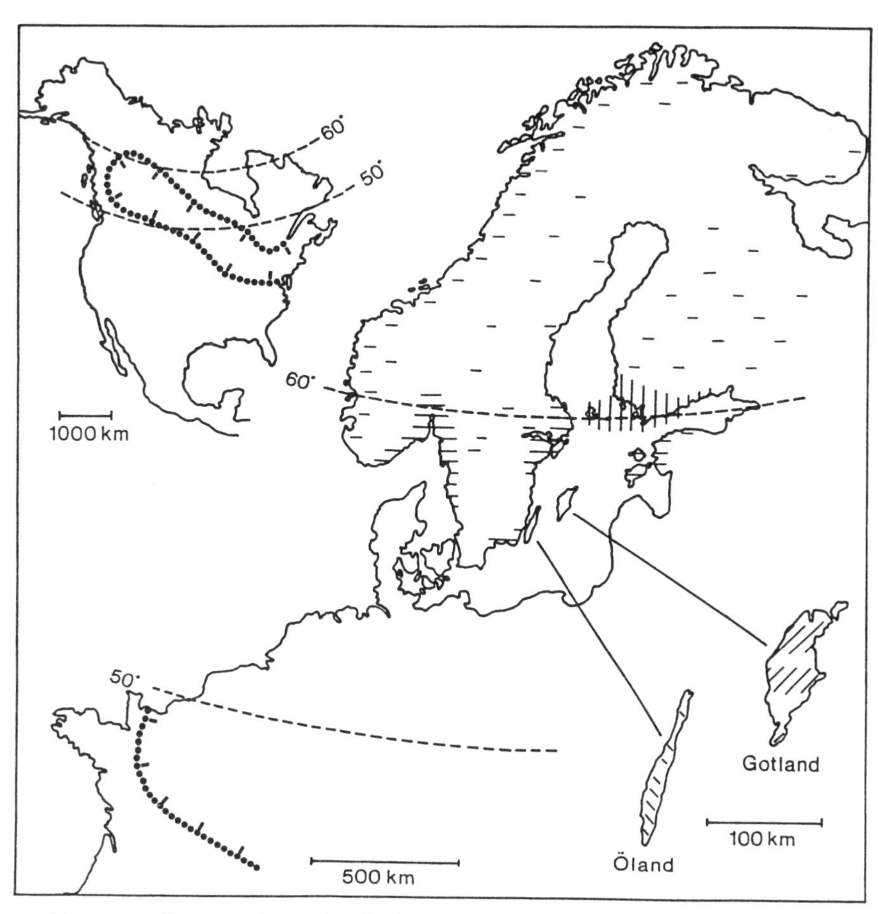

FIGURE 2. The approximate distribution of *C. tectorum* in Europe and North America (limit outlined by dotted lines). Hatched areas indicate the range of outcrop populations: horizontal lines = the mainland outcrop form; vertical lines = the S. Finland outcrop form; oblique lines = the outcrop forms on Öland (subsp. *pumila*) and Gotland. After Babcock (1947), Najda *et al* (1982) and Andersson (1990b, 1993a). Note different scales.

The process of differentiation has occurred without the formation of strong reproductive barriers (Andersson, 1993a) and must have been particularly rapid in the Baltic lowland area which emerged relatively late from the sea (6000-4000 BP). While analyses of variation in segregating F_2 progenies from hybridization between outcrop and weed populations suggest that extreme differences in leaf shape have a relatively simple genetic basis (Andersson, 1991a), other variable traits appear to be governed by multiple genetic factors, each having a relatively small effect on the phenotype (Table 1; Andersson, 1993a). Introgression of genes from other species is

TABLE 1. Distinct features of outcrop plants in different regions, possible selection pressure(s) responsible for their origin and the minimum number of segregating loci accounting for population differences (n_E; Wright, 1968; data from Andersson, 1991a, 1993a). Characters followed by the same letter denote a significant, positive rank association ($p < 0.001$) in the F_2 of crosses between contrasting populations (Andersson, 1993b). The mainland category refers to outcrop populations outside Öland, Gotland, and S. Finland (Fig. 1).

Trait	*Region*	*Selective Basis*	n_E
Leaf glands	Öland, Gotland	drought, insolation	
Early flowering	All regions	summer drought	
Dissected leaves	Öland, S. Finland	drought, overheating, nutrient deficiency	1 a
Short stature	All regions	drought, wind exposure, nutrient deficiency	3 abc
Small flowers	Gotland, mainland	few pollinators	6 b
Small seeds	Öland, Gotland, mainland	high fecundity, dispersability	3 c

unlikely, considering the strong interspecific crossing barriers and the lack of naturally occurring hybrids between *C. tectorum* and other species in this region (Babcock, 1947; Andersson, 1993a).

CHARACTER EVOLUTION IN *C. TECTORUM*

The monocarpic habit of *C. tectorum* probably evolved as a response to ephemeral or seasonally dry environments, since the closest relatives are perennial and occur in more mesic and stable habitats (Babcock, 1947). Direct and indirect evidence of a persistent seed bank (Andersson, 1990a) suggests that risk spreading is a further means of coping with environmental unpredictability, both on outcrops subject to occasional droughts, and in anthropogenic habitats subject to other kinds of disturbance (ploughing, road building, weed clearing, etc.).

Based upon comparative data obtained in two cultivation experiments (Andersson, 1989b,c), attempts were made to fit outcrop and weed populations of *C. tectorum* into the r-K continuum, which contrasts populations in habitats with various degrees of disturbance (MacArthur & Wilson, 1967). Contrary to predictions, weed populations combined the expected r-selected traits (rapid growth, a potential for a summer annual habit and a high seed yield) with more competitive traits (a tall stem and large seeds), a possible reflection of the relatively high levels of competition encountered by plants in many weed populations (Andersson, unpubl.). Coupled with experimental evidence that the trend towards autogamy has occurred independently of the trend towards a weedy habit (Andersson, 1989d), these observations are only partly congruent with predictions regarding the "ideal" weed (Baker, 1974).

The tall stem and the large seeds of the weed type also occur in more primitive species in the genus (Babcock, 1947) and probably evolved before *C. tectorum* spread into anthropogenic habitats. The wide and conspicuous flower heads characterizing the more or less autogamous weed populations (Andersson, 1989d) may also be ancestral, perhaps being a pleiotropic effect of the large overall size of plants in these populations (Table 1; Andersson, 1993b).

A more detailed hypothesis of the phylogenetic relationship between different races of *C. tectorum* was provided by a cladistic analysis of data obtained in a multivariate study of populations scattered throughout Europe and Canada (Andersson, 1993a). The analysis was hampered by the quantitative nature of most characters and the lack of outgroup data for some variables, and can only be used to reveal the major evolutionary trend(s) in the western part of the range. The single most parsimonious cladogram (Fig. 3) confirms the ancient morphology of most weed populations (particularly those outside northern Europe) and supports the notion that the colonization of rocky outcrops in the Baltic area has involved the greatest changes (Andersson, 1990b, 1991a, 1993a), including several "novel" traits not shared by the outgroup (Table 1), most of which parallel trends seen in the genus as a whole (Babcock, 1947).

One of the most extreme outcrop forms is subsp. *pumila* (Liljebl.) Sterner, a winter annual confined to thin, weathered soil on the calcareous grasslands ("alvars") on the Baltic island of Öland. Earlier taxonomic treatments (Sterner, 1938; Babcock, 1947), an extensive survey of herbarium material (Andersson, 1990b) and a series of common-garden experiments (Andersson, 1989a-d, 1991a-b, 1993a) have provided the following list of derived and presumably adaptive traits in subsp. *pumila*:

- compound leaves with small leaflets, contrasting with the weakly lobed leaves of weedy and non-weedy populations from more shady and fertile habitats. Apart from decreasing the "effective leaf size," thereby reducing transpiration and over-heating in a dry and sunny habitat, finely divided leaves also provide a more favourable leaf energy budget under nutrient-poor conditions (Givnish, 1987).
- a dense cover of glandular hairs on the leaves, contrasting with the more or less glabrous leaves of plants in weed populations and some outcrop popu-

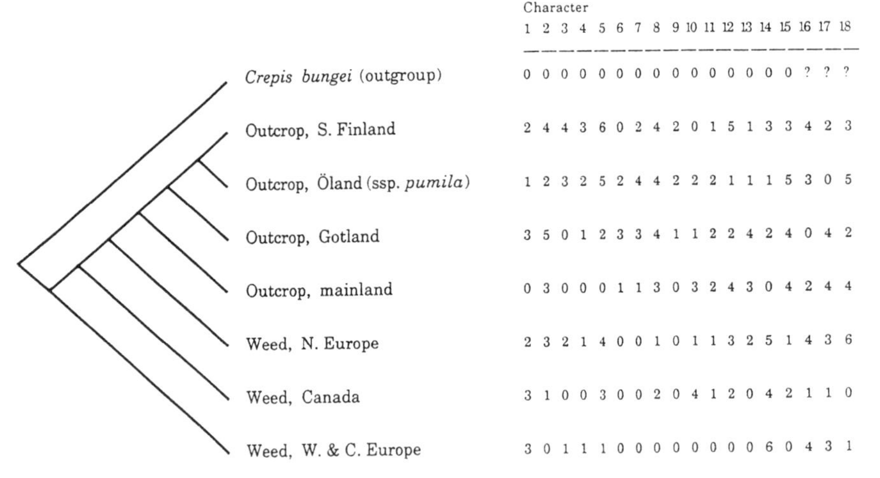

FIGURE 3. The minimum-length tree (PAUP, Swofford, 1990) connecting the major groupings of *C. tectorum* revealed in a numerical analysis of 52 European and Canadian populations (Andersson 1993a) and rooted with *C. bungei* (data from Babcock 1947). The consistency index was 0.59. All traits were treated as ordered multistates, the states being ranks from 0 (= closest to, or overlapping with, the outgroup) and weighted by 1/(n-1), where n is the number of states. Character states are shown to the right (ancestral-derived): (1) biennial-annual, (2) leaves large-small, (3) leaves oblanceolate-lanceolate, (4) leaves narrow-wide, (5) leaves dentate-dissected, (6) leaves glabrescent-glandular, (7) stem tall-short, (8) prop. of nodes with side branches low-high, (9) basal branches few-many, (10) side branches long-short, (11) involucre large-small, (12) flowers per head many-few, (13) ligules large-small, (14) styles yellow-dark green, (15) seeds large-small, (16) the presence or absence of anthocyanin leaf spots, (17) the degree of self-fertility and (18) the extent of seed dormancy.

lations. An adaptive role seems likely, given the strong association between leaf pubescence and the openness of the habitat found in other taxa (Johnson, 1975; Ehleringer & Clark, 1988).

- early flowering in the summer (May-June), contrasting with the later and more extended flowering of conspecific plants in more mesic weed habitats (June-September). The thin alvar soil is rapidly dried out in most summers, favouring plants which flower early and have rapid seed maturation (Widén, 1980; Andersson, 1988).
- a short, densely branched stem with more or less horizontal side branches, contrasting with the taller stature and more erect branches of plants from more closed habitats. The short stature of subsp. *pumila* and many other alvar plants (Witte, 1906; Pettersson, 1958) implies a selective advantage of decumbent growth in the exposed alvar habitat where plants suffer from wind damage, desiccation and nutrient deficiency. A similar conclusion was drawn by Turesson (1922) who found coastal populations of perennial plant species to be more prostrate in growth habit than conspecific populations in more mesic inland habitats.
- unusually small seeds, contrasting with the larger seeds of most weed

populations. Small propagules allow maternal plants to maintain a high fecundity (and dispersibility) in a habitat where seedlings experience relatively low levels of competition from other plants (Salisbury, 1942; Mazer, 1990), although a non-adaptive explanation is also possible (see below).

As comparative data provide indirect evidence for selection imposed by adverse local conditions, it seems useful to consider the distinct subsp. *pumila* as an ecotype *sensu* Turesson (1922); other forms of *C. tectorum* represent the ends of a more continuous clinal system (Andersson, 1990b, 1991a, 1993a) and show extensive differentiation among local populations (Andersson, 1991a), partly because of the greater potential for single individuals to establish new populations in these largely self-fertile forms (Andersson 1989d, 1993a). Characteristic *pumila* traits shared by the outgroup, or characters for which the ancestral status is uncertain (no outgroup data), include anthocyanic leaves, large and conspicuous flower heads (relative to other outcrop forms) and strong self-sterility (a unique feature in *C. tectorum*; Andersson, 1989d, 1993a), some of which may be ancestral rather than the result of local differentiation.

CURRENT PATTERNS OF SELECTION

A three-year demographic study (1986-1989) was conducted to identify targets of selection on morphology and life-history in a representative population of the morphologically derived subsp. *pumila* (Andersson, 1992). Most traits that were considered as "adaptive" in the comparative analyses (see above) were surprisingly variable in the study population and could be measured non-destructively. The main objective was to see whether individuals with distinctive *pumila* traits (deeply lobed leaves, early flowering, decumbent growth and small seeds) were favoured by selection and whether selection acted directly or as an indirect response due to selection on other, correlated traits. Following Lande and Arnold (1983), I assumed that phenotypic selection could be described as the statistical relationship between phenotypic variation and fitness, and used multiple regression to quantify the direct force of selection on each character (controlling for overall plant size and other traits).

In 1988 and 1989, the summer drought occurred too early in the season to have any selective effect on traits expressed after germination; no plant survived to the flowering stage. However, selection for imposing seed dormancy must have been strong before the two drought years, verifying the advantage of a between-year seed bank in this habitat (Andersson 1990a).

The selection regime in a season when a large number of plants survived to reproduction (1986-87) was only partly consistent with predictions based on comparative data (Table 2). Not surprisingly, overall plant size and major components of fecundity were positively correlated with life-time fitness, demonstrating strong selection on traits believed to be of general adaptive significance in most plants. The results also indicate a strong negative correlation between emergence time and life-time fitness, although the advantage of early emergence appears to be mediated through effects on subsequent leaf number or other characters expressed later in the ontogeny, as shown by the weak positive relationship between germination date and

TABLE 2. Patterns of past and current selection in subsp. *pumila*. The direction of past selection was inferred from comparative data (see text), while the current selection regime was revealed in a demographic study (Andersson, 1992). Net selection describes the change in the mean value of a character due to phenotypic selection and was calculated as the covariance between each of the phenotypic variables and relative fitness, while direct selection denotes the causal effect of the trait on fitness, all other traits held constant (multiple regression). Selection coefficients for pre-reproductive traits were obtained by summing the estimates based on differential viability and fecundity, while estimates of selection on architectural and reproductive characters express the relationship between each trait and fecundity. All estimates are expressed in standard deviation units. Sample sizes range from 511 to 1568 plants. Narrow-sense heritabilities (h^2) were calculated as 4 times the proportion of variance attributed to family in a comparison of 40 maternal sibships representing the same population (data from Andersson, 1991b). *** $p < 0.001$, ns = not significant, ne = not estimated.

		Current Selection		
Trait	*Past Selection*	*net*	*direct*	h^2
Germination time	?	-0.643***	+0.233***	ne
Leaf number	+	+1.998***	+1.674***	0.56-0.88***
Leaf dissection	+	+1.489***	+0.219***	0.60***
Flowering date	-	-0.558***	+0.050 ns	0.32***
Plant height	-	+0.398***	+0.184***	0.68***
Branch density	+	+0.899***	+0.381***	0.36***
Head number	+	+1.237***	ne	0.40***
Flowers per head	+	+1.106***	ne	0.36***
Seeds per ovule	+	+0.335***	ne	ne
Seed size	-	+0.399***	-0.020 ns	0.32***

fitness when adjusted for the other variables (multiple regression). While no tradeoff between seed size and fecundity could be detected, selection on other characteristic *pumila* traits generally acted in the predicted direction; plants with deeply lobed leaves survived better than those with weakly lobed leaves and a high branch density allowed vigorous plants to produce a large number of heads from a short stem. Moreover, apart from seed size and plant height, there was a relationship between fitness and the extent to which the plants approached the *pumila* phenotype, even though the selective advantage of early flowering disappeared when other traits were held constant. The direction of selection on plant height was opposite to that inferred from comparative data, with tall individuals being more fecund than those with a short stem, presumably because weather conditions were unusual in the summer of 1987 (frequent rains), a conclusion that also applies to the lack of direct selection on time of flowering. Given the long-term advantage of early flowering and decumbent growth in the alvar habitat (see above), my results indicate temporal fluctuations in selection on these characters (intermittent selection; Endler, 1986) and complement similar findings in other studies (Wolff, 1988; Galen, 1989; Scheiner, 1989; Van Tienderen & Van der Toorn, 1991) showing that ecologically differentiated populations may still be under selection.

A quantitative-genetic analysis based on 40 maternal sibships derived from the study population and raised under uniform greenhouse conditions (Andersson, 1991b) confirmed that most of the selected traits were heritable, suggesting that past selection has been too weak to eliminate the genetic variability in presumably adaptive characters and that the locally adapted subsp. *pumila* retains the ability to respond to changes in selection pressures (Falconer, 1981; Endler, 1986). The heritabilities (Table 2), calculated under the assumption that progenies within families were related as half-sibs, may be higher than those that determine the response to selection in the field, but there is evidence that at least one character is heritable under natural conditions: the size of the seeds produced by the plants sampled in the field was strongly correlated with the size of the seeds produced by their progeny in the greenhouse ($r = 0.53$, $p < 0.001$). When combined with comparative data on 11 populations grown in the same greenhouse (Andersson, 1991b), I found a strong positive correlation between the degree of population divergence in a trait and the extent of between-family variation (Table 3), implying stability in the relative variability for substantial periods of time. These observations are not compatible with the hypothesis that the population has reached equilibrium conditions of low genetic variance for characters connected with fitness (Falconer, 1981).

TABLE 3. The relationship between patterns of variability among and within populations, as measured by the Spearman rank correlation (r_S) between (1) the coefficient of variation (CV) of 11 population means and the CV of 40 family means representing a single population of subsp. *pumila*, (2) the character association (r_S) among the medians of 52 populations and among individual plants in two segregating F_2 families (n > 175), and (3) the CV of environment means (the amount of phenotypic plasticity) and the CV of overall population means (the degree of population divergence) in two experiments with populations replicated across the same sets of environments.

Parameter (# traits or correlations)	r_s	*Source*
1. Variation among populations and families (37 traits)	0.77	Andersson (1991b)
2. Trait associations, cross no. 1 (45 correlations)	0.80	Andersson (1993b)
2. Trait associations, cross no. 2 (21 correlations)	0.98	Andersson (1993b)
3. Plasticity vs population divergence, 1985 (35 traits)	0.84	Andersson (1989c)
3. Plasticity vs population divergence, 1986 (35 traits)	0.75	Andersson (1989c)

POPULATION DIVERGENCE IN PLASTICITY

Most studies of adaptation assume that selection acts on a single phenotypic expression of a trait. For instance, comparisons of ecological races are usually based on trait means expressed in a "standard" environment, whereas field estimates of selection generally describe the relationship between fitness and the "realized" phenotype of each individual. This approach is practical but ignores the possibility that selection can modify the norm of reaction, *i.e.*, the range of phenotypes that can be produced by a given genotype in response to a specific range of environmental conditions (Bradshaw, 1965; Via & Lande, 1985; Sultan, 1987). Hence, given the spatial variation in most habitats, more complex (and realistic) patterns of selection can probably be revealed if phenotypic plasticity is also considered.

To determine whether populations of *C. tectorum* differ in their response to different growth conditions, I extended the comparative analyses to a series of garden and greenhouse environments and used two-way analysis of variance and principal component analysis to compare the reaction norms. As regards the morphological characters, the plastic responses to contrasting environments usually involved several traits simultaneously in a pattern that appeared to be specific for each population, although different populations within the weed and outcrop forms tended to have more similar (multivariate) response patterns than other combinations of populations (Andersson, 1989c). A similar conclusion applies to some of the life-history characters; outcrop plants flower 2 or 3 weeks earlier than the weed plants when sown in the previous autumn, while the difference in flowering time is reversed for outcrop and weed plants that emerge in the spring (Andersson, 1989b). Taken together, these observations suggest that local selection pressures have favoured an integrated response to different growth conditions. However, the extent to which the reaction norm represents a target of selection in present-day populations remains unknown and will be the subject of further studies.

Some aspects of the morphology have been highly resistant to evolutionary change. All populations have retained a similar leaf shape at juvenile stages, despite the extensive divergence for leaf shape observed at adult stages (Andersson, 1989a), and the rank of characters with respect to their plasticity is similar in most populations, floral and shape variables being the least variable (Andersson, 1989c). Interestingly, the amount of response has not evolved independently of the mean; population divergence of overall character means has been greatest in the phenotypically most plastic characters and lowest in phenotypically stable characters (Table 3). Hence, available data from *C. tectorum* provide little evidence for a tradeoff between phenotypic plasticity and genetic specialization, or that high levels of plasticity have retarded evolutionary change by buffering the effects of natural selection (Sultan, 1987).

GENETIC CONSTRAINTS ON POPULATION DIFFERENTIATION

Inferences of past selection from comparative data can be misleading for traits that evolved in response to selective changes in correlated characters (Falconer, 1981; Endler, 1986). A quantitative-genetic study of a population of subsp. *pumila* (Andersson, 1991b) and analyses of character segregation in experimental populations derived from hybridization of widely different races of *C. tectorum* (Andersson, 1993b) have demonstrated consistent links between the size of vegetative and reproductive structures, some of which may have constrained population differentiation in this species. Firstly, I found head width to be positively genetically correlated with plant height ($r = 0.47$, $p < 0.001$) and seed size ($r = 0.72$, $p < 0.001$) in a comparison of maternal families representing a natural population of subsp. *pumila* (Andersson, 1991b, 1993b). Secondly, trait associations detected in the segregating progenies largely parallel those observed at the among-population level, providing a further connection between patterns of variability across taxonomic levels (Table 3).

One suggestion emerging from these studies is that the reduction in seed size associated with the colonization of rocky outcrops could be an allometric response to selection for a prostrate growth habit, rather than a means of increasing the fecundity in a habitat where large seed and seedling reserves are unnecessary; available evidence argue against a physiological tradeoff between seed number and seed size (Andersson, 1992, 1993b). Non-adaptive changes due to genetic correlations with plant height could also account for the small difference in the size of the attractive structures (flower and head size) between the decumbent, self-steriie and insect-pollinated subsp. *pumila* and the self-fertile plants of the much taller weed type (Andersson, 1989d, 1993a). Under the assumption that differences between present-day populations have originated as variation among genotypes within ancestral populations, it seems that underlying (genetic) constraints have indeed influenced large-scale patterns of differentiation in *C. tectorum.* Combined with the simple heritable basis of deeply lobed leaves (Andersson, 1991a) and the lack of genetically determined crossing barriers, despite extensive radiation in external morphology (Andersson, 1993a), quite small genetic changes could account for the origin of subsp. *pumila* and other outcrop forms in the Baltic region.

ACKNOWLEDGEMENTS

I thank B. Widén, H. Runemark and other colleagues for encouragement, discussions and critical reading of manuscripts. M. Widén made the illustration in Figure 1 and technical assistance were given by I. Larsson, K.-G. Forss, L. Herrosé, M. Christiansson and H. Persson. I wish to thank the staff of the Uppsala University Ecological Station at Ölands Skogsby for providing an excellent base for the field work. Linguistic advice by K. Ryde is also acknowledged. Parts of this study were supported by grants from the Harald E. Johansson Foundation. During the preparation of this paper, the author was supported by the Swedish Natural Science Research Council (post-doctoral fellowship).

References

Andersson, S. 1988. Limiting factors on seed production in *Crepis tectorum* ssp. *pumila. Acta Phytogeogr. Suec.* 76:9-20.

Andersson, S. 1989a. Variation in heteroblastic succession among populations of *Crepis tectorum. Nord. Jour. Bot.* 8:565-573.

Andersson, S. 1989b. Life-history variation in *Crepis tectorum* (Asteraceae). *Oecologia* (Berl.) 80:540-545.

Andersson, S. 1989c. Phenotypic plasticity in *Crepis tectorum* (Asteraceae). *Plant Syst. Evol.* 168:19-38.

Andersson, S. 1989d. The evolution of self-fertility in *Crepis tectorum* (Asteraceae). *Plant Syst. Evol.* 168:227-236.

Andersson, S. 1990a. The relationship between seed dormancy, seed size and weediness in *Crepis tectorum* (Asteraceae). *Oecologia* (Berl.) 83:277-280.

Andersson, S. 1990b. A phenetic study of *Crepis tectorum* in Fennoscandia and Estonia. *Nord. Jour. Bot.* 9:589-600.

Andersson, S. 1991a. Geographical variation and genetic analysis of leaf shape in *Crepis tectorum* (Asteraceae). *Plant Syst. Evol.* 178:247-258.

Andersson, S. 1991b. Quantitative genetic variation in a population of *Crepis tectorum* ssp. *pumila* (Asteraceae). *Biol. Jour. Linn. Soc.* 44:381-393.

Andersson, S. 1992. Phenotypic selection in a population of *Crepis tectorum* ssp. *pumila* (Asteraceae). *Canadian Jour. Bot.* 70:89-95.

Andersson, S. 1993a. Morphometric differentiation, patterns of interfertility and the genetic basis of character evolution in *Crepis tectorum* (Asteraceae). *Plant Syst. Evol.* 184:27-40.

Andersson, S. 1993b. Population differentiation in *Crepis tectorum* (Asteraceae): patterns of correlation among characters. *Biol. Jour. Linn. Soc.* 49:185-194.

Babcock, E. B. 1947. The Genus *Crepis I-II.* Univ. California Press, Berkeley, Los Angeles.

Baker, H. G. 1974. The evolution of weeds. *Annu. Rev. Ecol. Syst.* 5: 1-24.

Bradshaw, A. D. 1965. Evolutionary significance of phenotypic plasticity in plants. *Adv. Genet.* 13:115-155.

Clausen, J., D. D. Keck and W. H. Hiesey. 1940. Experimental studies on the nature of species. I. The effect of varied environments on Western North American plants. *Carnegie Inst. Washington, Publ.* 520. 450 pp.

Darwin, C. 1859. *The Origin of Species by Means of Natural Selection.* Murray, London, UK.

Ehleringer, J. R. and C. Clark. 1988. Evolution and adaptation in *Encelia* (Asteraceae). Pages 221-248 *in* L. D. Gottlieb & S. K. Jain, eds., *Plant Evolutionary Biology.* Chapman and Hall, New York.

Endler, J. A. 1986. *Natural Selection in the Wild.* Princeton Univ. Press, Princeton, New Jersey.

Falconer, D. S. 1981. *Introduction to Quantitative Genetics.* Longman, London.

Felsenstein, J. 1988. Phylogenies and quantitative characters. *Ann. Rev. Ecol. Syst.* 19:445-471.

Galen, C. 1989. Measuring pollinator-mediated selection on morphometric traits: Bumblebees and the alpine sky pilot, *Polemonium viscosum. Evolution* 43:882-890.

Givnish, T. J. 1987. Comparative studies of leaf form: Assessing the relative roles of selective pressures and phylogenetic constraints. *New Phytol.* 106(suppl.):131-160.

Johnson, H. B. 1975. Plant pubescence: An ecological perspective. *Bot. Rev.* 41:233-258.

Jordan, N. 1991. Multivariate analysis of selection in experimental populations derived from hybridization of two ecotypes of the annual plant, *Diodia teres* W. (Rubiaceae). *Evolution* 45:1760-1772.

Lande, R. and S. J. Arnold. 1983. The measurement of selection on correlated characters. *Evolution* 37:1210-1226.

MacArthur, R. H. and E. O. Wilson. 1967. *The Theory of Island Biogeography.* Princeton Univ. Press, Princeton, New Jersey.

Mazer, S. J. 1990. Seed mass of Indiana Dune genera and families: Taxonomic and ecological correlates. *Evol. Ecol.* 4:326-357.

Najda, H. G., A. L. Darwent and G. Hamilton. 1982. The biology of

Canadian weeds 54. *Crepis tectorum. Canadian Jour. Plant. Sci.* 62:473-481.

Pettersson, B. 1958. Dynamik och konstans i Gotlands flora och vegetation. *Acta phytogeogr. suec.* 40:1-288.

Primack, R. B. and H. Kang. 1989. Measuring fitness and selection in wild plant populations. *Ann. Rev. Ecol. Syst.* 20:367-396.

Salisbury, E. J. 1942. *The Reproductive Capacity of Plants.* Bell, London. UK.

Scheiner, S. M. 1989. Variable selection along a successional gradient. *Evolution* 43:548-562.

Sokal, R. R. 1978. Population differentiation: Something new or more of the same? Pages 215-239 *in* P. F. Brussard, ed., *Ecological Genetics. The Interface.* Springer-Verlag, New York.

Sterner, R. 1938. Flora der Insel Öland. *Acta phytogeogr. suec.* 9: 1-169.

Sultan, S. E. 1987. Evolutionary implications of phenotypic plasticity in plants. *Evol. Biol.* 21:127-178.

Swofford, D. L. 1990. *PAUP. Phylogenetic Analysis Using Parsimony, Version 3.0.* Computer package. Champaign, Illinois: Illinois Nat. Hist. Surv.

Turesson, G. 1922. The genotypical response of the plant species to the habitat. *Hereditas* 3:211-350.

Van Tienderen, P. H. and J. Van der Toorn. 1991. Genetic differentiation between populations of *Plantago lanceolata.* II. Phenotypic selection in a transplant experiment in three contrasting habitats. *Jour. Ecol.* 79:43-59.

Venable, D. L. 1984. Using intraspecific variation to study the ecological significance and evolution of plant life-histories. Pages 166-187 *in* R. Dirzo & J. Sarukhan, eds., *Perspectives on Plant Population Ecology.* Sunderland, Mass.

Via, S. and R. Lande. 1985. Genotype-environment interaction and the evolution of phenotypic plasticity. *Evolution* 39:505-522.

Wanntorp, H.-E. 1983. Historical constraints in adaptation theory: Traits and non-traits. *Oikos* 41:157-160.

Widén, B. 1980. Flowering strategies in the *Helianthemum oelandicum* (Cistaceae) complex on Öland, Sweden. *Bot. Notiser* 133:99-115.

Witte, H. 1906. Till de svenska alfvarväxternas ekologi. *Thesis.* Uppsala Univ.

Wolff, K. 1988. Genetic analysis of ecologically relevant morphological variability in *Plantago lanceolata* L. *Theor. Appl. Genet.* 75:772-778.

Wright, S. 1968. *Evolution and the Genetics of Populations. I. Genetics and Biometrical Foundations.* Univ. Chicago Press, Chicago, Illinois.

Ecological Adaptation of Cereal Rusts

James Mac Key
Department of Plant Breeding
Swedish University of Agricultural Sciences
Uppsala, Sweden

Turesson (1922, 1925, 1931) made us understand that environment does not interact only with the genetic constitution of an organism. Over time and on a population basis, environment will also influence the genetic composition in such a way that the interaction becomes even more effective and expedient. He demonstrated this important observation on evolution and differentiation mainly by observing how plant species adapt themselves to different habitats by adjusting their morphological and physiological traits. The present contribution will show an outcome when environments favor different types of reproduction and their consequences. This time mycological objects are chosen, which especially for the young Göte Turesson were far from unfamiliar.

REPRODUCTION PATTERN OF RUSTS IN DIFFERENT REGIONS

Cereal rusts (*Puccinia* spp.) are obligate parasites specialized on annual plants, *i.e.*, hosts that normally are not available the whole year around. For survival they must find a way to bridge the unfavourable season. They have evolved an alternative host system mediated by a sexual phase. As man has spread the cereal grasses over large areas around the world, the rusts have adapted their mode of reproduction to the prevailing geographical situation.

Kenya and adjacent regions in East Africa may be mentioned as an area where it is possible to plant wheat at any month of the year. The rusts can easily survive through the asexually propagated uredospores all the time. America offers a situation where uredospores are able to be transported by winds from North American summer crops to Central American winter crops of the same species and vice versa. This move to and fro, described as the Puccinia Path (Frey *et al.*, 1977), will also favor the more opportunistic asexual propagation, where the rusts do not utilize their complete reproduction cycle. A similar situation is met on both sides of the Straits of Gibraltar. In Australia, where the alternate hosts to stem and leaf rusts are missing (Watson & Luig, 1968), survival must for another reason also rely on uredospores alone.

Scandinavia may be taken as the other extreme, where the complete reproduction cycle is apparently a must. The cold winters do not allow uredospores to survive, and the wide waters around seldom allow wind transport of spores from the continent. The last time such an event happened and caused an earlier and thus devastating epidemic of wheat stem rust was in 1951 (Mac Key, 1952) Such rare invasions are, however, not able to survive and establish themselves because races invading from the south are too slow in shifting from uredo- to teliostage for safe overwintering.

The impact of the asexual reproduction based on the uredineal stage alone versus the need for a complete cycle including sexual recombination may be illustrated by comparing the virulence pattern of rusts in the United States and Sweden. The choice of these two countries can be said to have a special meaning in a publication in honor of Göte Turesson.

RACE PATTERN OF RUSTS GOVERNED BY MODE OF REPRODUCTION

Before entering the demonstration it might, however, be appropriate to remember that the race-specific resistance pattern, developed in most obligate parasitisms, offers a unique analytical precision. In this case, there exists a direct gene-for-gene relation between the avirulence/virulence of the parasite and the resistance/susceptibility of the host (Flor, 1942). A specific gene for virulence of the parasite is able to overcome a specific gene of resistance of the host but no other (Person, 1959). The precision is like a key that only fits to a certain lock. This interrelation allows us to identify the existence of a certain virulence gene by use of the matching gene for resistance. We must know which gene for resistance we work with, and in order to avoid confusion, this gene must occur alone in an otherwise susceptible genetic background.

The old race key for oat stem rust (*Puccinia graminis Pers. f. sp. avenae* E. et H.) happens to fit this prerequisite, and this rust will thus be used as a first example. The race key comprises four differentials which later have been proved to carry one dominant gene for race-specific resistance each (*cf.* Mac Key, 1977). Originally denoted A, *B*, *D* and *E*, they can thus by simplicity be said to be overcome by the recessive and corresponding virulence genes *a, b, d* and *e*.

The races thus identified were numbered, as they were found. Altogether 26 different races were recorded in North America and 16 in Europe. The 10 more races recorded in North American are all distinguished by having a mesothetic reaction. This mixed and temperature-unstable syndrome could later be shown to be caused by a superimposed plasmic effect (Green & McKenzie, 1967) and should in a deciphered system be read as susceptible. As expected four genes can occur in 16 different recombinations as shown in Table 1. The above-mentioned finding explains why two numbered races are represented by one true virulence gene in some combinations.

TABLE 1. Race Spectra of Oat Stem Rust in Sweden and U.S.A. in 1956-1959 (Mac Key, 1977).

		Relative prevalence (%) of oat stem rust race number/pathotype:																	
Year	n	1	11	1A	3	2,5	11A	4	8,10	3A	2A,5A	7,12	4A	8A	10A6,13	7A,12A	6A,13A
		—	*a—*	*-b—*	*—d-*	*—e*	*ab—*	*a-d-*	*a--e*	*-bd-*	*-b-e*	*—de*	*abd-*	*ab-e*	*a-de*	*-bde*	*abde*
Sweden:																	
1956	32	–	–	–	31	–	–	28	–	6	–	13	–	–	13	3	6
1957	61	–	–	–	7	–	–	7	–	3	–	11	3	–	13	25	31
1958	96	–	–	–	24	–	–	14	–	5	–	20	3	–	15	6	14
1959	133	–	–	–	12	–	–	9	–	7	–	16	5	–	10	21	20
1956-59	322	–	–	–	16	–	–	12	–	6	–	16	4	–	12	15	19
U. S. A.:																	
1956	476	–	–	–	–	16	–	–	15	–	–	66	–	–	1	2	–
1957	522	–	–	–	–	12	–	–	21	–	0	59	–	–	2	6	0
1958	286	–	–	–	–	14	–	–	26	–	–	54	–	–	1	5	–
1959	230	–	–	–	–	7	–	–	11	–	–	59	1	–	10	10	2
1956-59	1514	–	–	–	–	13	–	–	19	–	0	60	0	–	3	5	0

The years 1956-59 are chosen, since race inventories from this period are the only ones available from Sweden. In the same period, the resistance gene *B* happened to be introduced in North America and gives thus a good illustration of the effect of a host selection pressure. Conscious resistance breeding introducing genes *A* and *D* started already in the 1940s in North America, while no oats carrying any of the four genes of race-specific resistance have ever been grown commercially in Scandinavia.

Having in mind that oat breeders have challenged the oat stem rust to develop races with virulence gene *a, d* and later *b*, the American race pattern show no tendency of "unnecessary" virulence accumulation. This is, however, what the Swedish spectrum does in spite of no host selection pressure at all. A long prevailing idea that genes for virulence have a negative general fitness (Van der Plank, 1968) offering a balanced polymorphism or stabilizing selection, thus fits only for the pattern in U.S.A.

Another observation can be made that genes for virulence may have pleiotropic effects. This again influences their prevalence irrespective of a host selection pressure. Virulence gene *d* is represented in all Swedish samples. Gene *e*, not challenged by the oat breeders, is represented in all samples but one in U.S.A., where barberry bushes also do exist. Apparently the two genes must have different ecological significance in the two regions.

The very same picture can be found by studying e.g. wheat brown rust (*Puccinia recondita Rob. ex Desm. f.sp. tritici*). For consistency, the demonstration is again limited to four gene pairs and the same years. In Table 2, Portugal is also included in order to illustrate the situation in that region of Europe. Again, no host selection pressure is influencing race pattern in Sweden, nor in Portugal. U.S.A. had at that time introduced three of the concerned genes for resistance. The concept of stabilizing selection is supported by the spectrum found in Portugal as well as in U.S.A., in the latter case considering the host selection pressure already applied. However, in Sweden wheat leaf rust again shows a clear tendency of an "unnecessary" gene accumulation, which in fact characterizes the whole of eastern Europe.

TABLE 2. Race Spectra of Wheat Leaf Rust in Sweden, Portugal and U.S.A. in 1956-1959 (Mac Key, 1981).

		Relative prevalence (%) of wheat leaf rust pathotype:															
Year	n	—	*a—*	*-b–*	*–c-*	*—d*	*ab–*	*a-c-*	*a–d*	*-bc-*	*-b-d*	*–cd*	*abc-*	*ab-d*	*a-cd*	*-bcd*	*abcd*
Sweden:																	
1956	189	–	–	–	–	–	–	–	–	–	–	–	–	39	–	–	61
1957	84	–	–	–	–	–	–	–	–	–	–	–	–	21	–	47	32
1958	158	4	–	–	–	7	–	–	–	–	51	3	–	4	–	30	1
1959	154	2	–	–	–	4	–	–	–	–	44	5	–	14	–	31	–
1956-59	585	2	–	–	–	3	–	–	–	–	25	2	–	20	–	23	25
Portugal:																	
1956	65	43	–	11	–	–	–	–	–	43	–	3	–	–	–	–	–
1957	70	43	–	–	–	–	–	–	–	57	–	–	–	–	–	–	–
1958	119	49	–	27	–	–	–	–	–	23	–	1	–	–	–	–	–
1959	231	56	–	10	7	–	–	–	–	22	1	–	–	–	–	4	–
1956-59	485	50	–	13	3	–	–	–	–	30	1	1	–	–	–	2	–
U. S. A.:																	
1956	674	5	1	–	24	–	6	28	–	–	–	1	26	0	8	–	1
1957	997	2	–	–	42	–	3	26	–	0	–	1	16	2	7	–	1
1958	935	11	–	–	32	2	7	16	–	–	–	7	12	4	8	–	1
1959	1447	1	–	–	53	0	4	10	–	0	–	11	11	3	3	–	4
1956-59	4053	5	0	–	40	0	5	19	–	0	–	6	15	2	6	–	2

Symbol key: *a* = virulence gene matching *Lr-1* in Malakof,
b = " " " *Lr-2a* in Webster,
c = " " " *Lr-3* in Mediterranean/Democrat,
d = " " " *Lr-11* in Hussar.

Though somewhat less clear, virulence gene *d* (matching resistance gene *Lr11*) seems to have a supporting function in Sweden, gene *b* (*Lr2a*) in Portugal and gene *c* (*Lr3*) in U.S.A.

We can learn from the two examples that host:pathogen relationships in an obligate parasitism cannot be based on host selection pressure alone and thus described by simple mathematical models (*cf.* Groth & Person, 1977; Leonard, 1977; Marshall & Pryor, 1978).

Genes for virulence may obviously have pleiotropic effects acting on their general fitness or having a supporting function, which influences their relative prevalence. The specific pattern varies with geographical region, *i.e.*, with large-scale ecological niche.

As with most other genes, those governing virulence are also dependent on their genetic background for their general fitness. The specific and general gene erosion in a pathogen population submitted to the assortative effect of a race-specific host selection pressure is to a considerable extent governed by existing chances for recombinations.

In addition, it can also be stated that host alternation for safe survival favors a genetic system able to store genes used only seasonally. It might also be mentioned that the dikaryotic constitution of the rust uredospores is an important prerequisite for preadaptation, not found *e.g.* in connection with the true haploid cereal powdery mildews, *Erysiphe* spp. In contrast to rust, this pathogen follows the pattern of stabilizing selection in Sweden as well as apparently everywhere (Mac Key, 1981).

ADAPTATION OF BREEDING PROGRAMS FOR RACE-SPECIFIC RESISTANCE

Obviously, the adaptive as well as the preadaptive pattern of rust populations are highly dependent on degree of genetic isolation, not only geographically but also on how they commonly reproduce. The different strategies must be understood for a proper planning of a breeding program aiming at a race-specific resistance.

Modern proposals for breeding towards race-specific resistance (Table 3) can be separated into two fundamentally different approaches, one aiming at gene accumulation, the other at gene distribution, or in other words often used, pyramidizing versus deployment.

Within each group, one intention is to find a *modus vivendi* with

TABLE 3. Strategies of Breeding for Race-specific Resistance in Plants.

A. APPROACH THROUGH SYNTHESIS		
1.	Breeding cultivars with race-specific resistance against a given pathogen by using a	
	a..	complete protection.
	b.	reduplicated complete protection.
2..	Breeding of cultivars with complete, moderate race -specific resistance against a given pathogen population with the intention uniformly to slow down selection pressure on the pathogen.	
B. APPROACH THROUGH DIVERSIFICATION IN SPACE		
1.	Construction of multi-line cultivars or varietal blends with effective genes for race-specific resistance distributed to separate components.	
2.	Coordinated use of different genes or sets of genes for race-specific resistance over geographic zones covering a common pathway of the pathogen or for cultivars of the same species sown, *e.g.*, in autumn versus spring.	
C. APPROACH THROUGH DIVERSIFICATION IN TIME		
1.	Cyclic shift in using individual genes for race-specific resistance known to promote stabilizing selection in the pathogen population (*cf.* Fig. 1).	

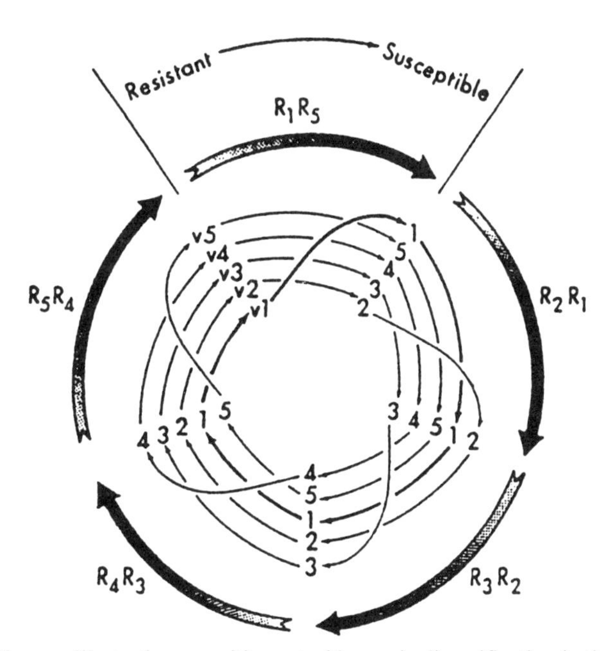

FIGURE 1. Diagram illustrating parasitic control by genic diversification in time. R_1 - R_5 in the outer circle stand for race-specific resistance genes in the host cultivar(s). These genes are being continuously exchanged through a rolling system as soon as the matching virulence genes v_1 - v_5 in the pathogen population become too frequent. The five inner circuits are intended to describe the changing frequencies of the v genes under a process of stabilizing selection (Person 1966).

the parasite on a level acceptable for yield but perhaps never for quality if aesthetic evaluations intervene. The use of complete but moderate protection and the multi-line concept are examples of this approach. They are similar in end effect. In this respect they have also some similarity with nonspecific resistance. They all slow down total infection potential without major disturbances of the racial balance. The mechanisms behind are, however, fundamentally different. The complete but moderate resistance strategy is as nonspecific resistance applicable against pathogen populations able to build up complex races, or zooparasites able to select their prey. The multi-line concept presupposes stabilizing selection to regulate the racial pattern. Vital races will be at least predominantly those matching each of the components of the blend. Since spores will land at random, they will only be able to infect the matching component. This will imply a reduction of the epidemic potential of the parasite population in proportion to the number of different components of the host population.

In the other category, the parasite is more directly challenged. The approach through more or less complete protection against a threatening pathogen population is an attempt to first completely conquer and secondly drastically erode the overall gene pool of the pathogen, and thus limit its evolutionary recovery. It is important that this stroke is made all at once, especially if the pathogen population shows high ability for gene accumulation. The diversification strategy by challenge presupposes a stabilizing selection. If so, it will assure that the pathogen has to run an everlasting relay-race, programmed in advance by the breeders rather than a breeder alone.

The proper choice of method is thus not unimportant. Parasite populations able to form and maintain complex races by themselves are difficult to defeat by any kind of deployment strategy. Parasite populations restricted to stabilizing selection can be handled by any of the methods discussed. Whenever genic flexibility of this category is high (*cf. e.g.*, cereal powdery mildews, Mac Key, 1981), diversification has some preference. It is also here less disturbing that diversification of the host automatically implies diversification of its parasite, giving it an improved evolutionary potential.

It is important that all breeders within a certain region are aware of the fact that breeding for race-specific resistance by diversification or by gene accumulation must never occur simultaneously with the same set of resistance genes within a natural epidemic region of a parasite. Increased competition rather than cooperation due to commercialization of plant breeding makes this principle, however, difficult to apply in practice.

SUMMARY

Dependent on regional circumstances, cereal rusts may reproduce either asexually or via host alternation mediated by a sexual phase. This difference in recombinatory capacity and thus ability to adjust genetic background seem to have important consequences for adaptation as well as preadaptation to a changing host resistance pattern. The general fitness, *i.e.* the prevalence of race-specific genes for virulence independent on host selection pressure, may differ from one environment to another. Oat stem rust and wheat leaf rust are used as demonstrations of these observations.

Literature Cited

Flor, H. H. 1942. Inheritance of pathogenicity in *Melampsora lini*. *Phytopathology* 32:653-669.

Frey, K. J., Browning, J. A. and Simons, M. D. 1977. Management systems for host genes to control disease loss. *Ann. New York Acad. Sci.* 287:255-274.

Green, G. J. and McKenzie, R. I. H. 1967. Mendelian and extrachromosomal inheritance of virulence in *Puccinia graminis* f.sp. *avenae*. *Canadian Jour. Genet. Cytol.* 9:785-793.

Groth, J. V. and Person, C. O. 1977. Genetic interdependence of host and parasite in epidemics. *Ann. New York Acad. Sci.* 287:97-106.

Leonard, K. J. 1977. Selection pressures and plant pathogens. - *Ann. New York Acad. Sci.* 287:207-222.

Mac Key, J. 1952. Svartrosten och veteförädlingen (The stem rust and the breeding of wheat). *Lantbruksveckan, Stockholm* 1952:205-221.

Mac Key, J. 1977. Strategies of race-specific phytoparasitism and its control by plant breeding. *Genetica, Acta Biol. Jugoslav.* 9:237-255.

Mac Key, J. 1981. Alternative strategies in fungal race-specific parasitism. *Theor. Appl. Genet.* 59:381-390.

Marshall, D. R.and Pryor, A. J. 1978. Multi-line varieties and disease control. I. The 'dirty crop' approach with each component carry-

ing a unique single resistance gene. *Theor. Appl. Genet.* 51:177-184.
Person, C. 1959. Gene-for-gene relationships in host:parasite systems. *Canadian Jour. Bot.* 37:1101-1130.
Person, C. 1966. Genetic polymorphism in parasitic systems. *Nature* 212:266-267.
Turesson, G. 1922. The genotypic response of the plant species to the habitat. *Hereditas* 3:211-350.
Turesson, G. 1925. The plant species in relation to habitat and climate. *Hereditas* 9:81-101.
Turesson, G. 1931. The selective effect of climate upon plant species. *Hereditas* 15:99-152.
Van der Plank, J. E. 1968. *Disease Resistance in Plants.* Academic Press, New York. 349 pp.
Watson, I. A. and Luig, N. H. 1968. The ecology and genetics of host-pathogen relationships in wheat rust in Australia. Pages 227-238 *in Proc. 3rd Internat. Wheat Genet. Symp.*, Canberra 1968. Australian Acad. Sci., Canberra.

Limits to Adaptive Population Differentiation of Quantitative Traits in Plants

Gerrit A. J. Platenkamp[1] and Ruth G. Shaw[2]

[1] Jones & Stokes Associates, Inc., Environmental Planning and Natural Resource Sciences, 2600 V Street, Sacramento, CA 95818; [2] Department of Ecology, Evolution, and Behavior, University of Minnesota, St. Paul, MN 55108.

From Turesson (1922, 1925) to the present, numerous well-documented examples of the differentiation of species into genetically distinct populations have been reported (Heslop-Harrison, 1964; Langlet, 1971; Hamrick, 1982). In many instances the case can be made that such differentiation has been the result of natural selection and that at least some of the traits by which the populations are differentiated represent adaptations (*e.g.*, Clausen, *et al.*, 1948; Gurevitch, 1992a). In some cases differentiation has been extremely rapid, on the order of a few generations (Wu, *et al.*, 1975; Snaydon, 1970), and has been associated with large selection differentials (Jain & Bradshaw, 1966; Davies & Snaydon, 1976). Abundant evidence supports the notion that natural selection can be very potent in effecting evolutionary change.

Unfortunately, it is much easier to observe the end results of evolution than the process itself. Too often reckless inferences of evolutionary processes have been made from the observed results of evolution only. This is exemplified by the "adaptive story-telling" that was so eloquently criticized by Gould and Lewontin (1979). They advocated the study of constraints on evolution as an alternative to the "adaptationist programme." Since the publication of their paper, limits on evolution have become an increasingly important focus of research, to the point that — according to some — we are "bombarded with all kinds of possible constraints to the process of natural selection" (Antonovics & Van Tienderen, 1991). Antonovics & Van Tienderen (1991) were critical of the myriad of different meanings carelessly given to the term "constraint" in the evolutionary literature. A further complaint is that the fitness differential associated with the traits under study is taken for granted or is assumed from indirect sources in most studies that claim constraints on selection. Perrin & Travis (1992) point out that too often constraints on evolution are invoked *ad hoc*, as untested hypotheses to explain observed patterns, while experiments to test these hypotheses are often not done. This use of the concept of constraints offers little improvement over the "adaptationist" programme.

In this chapter we will argue for an experimental approach to quantifying limits to evolution by natural selection. We believe that carefully designed field experiments that combine quantitative genetics and experimental demography can offer an approach to the study of limits to natural selection that avoids many of the aforementioned problems. We use "limits" to natural selection rather than "constraints," and define those as any "factors that prevent or reduce the potential for natural selection to result in the most direct ascent of the mean phenotype to an optimum" (Loeschcke, 1987:1). We acknowledge that other evolutionary forces, such as genetic drift, gene flow and mutation may be involved, but discussion of such limits to evolution are largely outside the scope of this chapter, where we discuss evolution by natural selection. Our discussion will focus on higher plants, but similar approaches have been applied successfully to animals (*e.g.*, Via, 1991; Rawson & Hilbish, 1991). First, we will develop the basic concept behind the experimental approach. Then, we will discuss a quantitative genetic approach to estimating components of phenotypic variation expressed in field experiments. Next, we will use examples of our own work to demonstrate the use of these techniques in studying limits on natural selection. Lastly, we will briefly examine the plant literature for evidence of limits on the evolutionary response to natural selection of quantitative traits.

EXPERIMENTAL EVOLUTIONARY DEMOGRAPHY

In 1924, Clements and Goldsmith introduced the use of the "phytometer," based on the important concept that plant responses can be used as integrated measures of those aspects of the environment that are relevant to the plant. They exposed replicates of potted plants or sods to different environments and recorded their responses (*e.g.*, growth or water use). Clausen, *et al.* (1940) developed a genetic approach to the use of phytometers in a series of classic transplant experiments. They demonstrated that plant responses in nature can be used to investigate the relative influences of environment and genetic background on the phenotype. For example, they transplanted material of populations of *Achillea millefolium* originating from different elevations to common gardens at a range of elevations in the Sierra Nevada (Clausen, *et al.*, 1948; see also Gurevitch, 1992a, 1992b). They found that in spite of intraspecific variation in growth, populations had highest relative fitness when grown at the elevation from which they originated. In their experiments plants were grown in spaced trials. Environmental variation due to micro-spatial variation within transplant sites (*e.g.*, neighboring plants) was therefore minimized. Consequently, the potential differences in plant performance among sites were likely accentuated, at the expense of ecologically realistic measurements of fitness.

As plant demography became a focus for research (Harper & White, 1974; Harper, 1977) the phytometer approach was refined to that of the "life-history phytometer" (Fowler & Antonovics, 1981; Antonovics & Primack, 1982). In this method genetically related seeds, seedlings or cuttings (clones) are transplanted into different environments. The survivorship and reproductive schedules of the transplants are used as measures of the environment (see Antonovics, *et al.*, 1988). Transplants can be made either into unaltered natural environments (*e.g.*, Antonovics & Primack, 1982; Schmidt & Levin, 1985), or into experimentally altered natural environments (Kelley & Clay, 1987; Platenkamp & Foin, 1990), while the level of genetic detail varies from inter-population to clonal variation. This approach has revealed striking examples of prior genetic divergence among populations and genetic variation within populations. Because much of this work has lacked genetic detail, its use for predicting future evolution is limited. The value of this approach can be expanded to permit the estimation of an upper bound of genetic variation available for future evolutionary response with a refinement of the genetic design, as we will advocate below.

SOURCES OF LIMITS TO THE EVOLUTION OF QUANTITATIVE TRAITS

Before we can discuss our approach to limits on evolution by natural selection of quantitative traits, we first need to examine the important parameters that determine the rate of an evolutionary response to selection.

The response to selection of a trait (R) depends on its heritability (h^2), and the selection differential (S, the difference between the mean trait value of the selected parents and the mean of the population), applied to that trait:

$$R = h^2 S \,, \qquad (1)$$

where

$$h^2 = V_A / V_P \,, \qquad (2)$$

V_A is the additive genetic variance and V_P is the phenotypic variance (Falconer, 1989). The phenotypic variance is the sum of a number of variance components, but assuming no dominance, non-nuclear parental effects, epistasis or genotype-environment interaction we can write

$$V_P = V_A + V_E \,, \qquad (3)$$

where V_E is the environmental variance. In nature this component is often larger than the genetic causes of phenotypic variation combined (Platenkamp & Shaw, 1992). So, both the magnitude of additive genetic variation and the magnitude of environmental variation are important in determining the rate of response to natural selection. There will be no selection response in a trait if there is no genetic variation in that trait ($V_A = 0$). Selection response is expected to proceed slowly when additive variance is non-zero, but the environmental variation is much larger than the additive genetic variation ($V_E >> V_A$). In practice evolution may stall when very slow selection response due to very low heritability is opposed by migration or rapid environmental change.

When at least one of several correlated traits is under selection, the joint response of all traits considered is predicted by:

$$\Delta\bar{z} = GP^{-1}s, \qquad (4)$$

and

$$P = G + E, \qquad (5)$$

respectively, where P, G and E are now the matrices of phenotypic, additive genetic, and environmental variances and covariances, and $\Delta\bar{z}$ and s are the vectors of response and selection differentials, respectively, for the traits involved (Lande, 1979). From (4) and (5) it is clear that not only the genetic and environmental variances of traits are important for predicting selection response but also their covariances (and thus correlations). While environmental covariances between two traits that are both positively correlated with fitness are usually positive (Reznick, *et al.*, 1986), genetic covariances may be negative, such that selection for an increase in one of the traits leads to a correlated negative response to selection in the other trait, due to pleiotropy or linkage. Such adverse genetic correlations may represent an important factor limiting the combinations of character states likely to arise by natural selection (Gould & Lewontin, 1979; Maynard Smith, 1983).

Once we consider not only the evolution of the overall population mean phenotype, but also its response to environmental variation (*i.e.*, phenotypic plasticity), the genotype-environment interaction variance needs to be taken into account. Equation (5) should now be changed to

$$P = G + E + I , \qquad (6)$$

where I is the matrix of variances and covariances between traits due to genotype-environment interaction (Platenkamp & Shaw, 1992). The genotype-environment interaction variance can be decomposed into a component due to differences in genetic variance within the different environments, and a component due to genetic correlation between traits expressed in different environments (Cockerham, 1963). This latter component (henceforth, "crossing-interaction variance") is particularly relevant to the study of population differentiation and the evolution of eco-geographical races because it represents the potential for the evolution of specialization of genotypes to different environments (Via & Lande, 1985; Gillespie & Turelli, 1989). A multivariate approach to both an organism's traits and its environment can overcome, in theory (if not fully in practice), one of the main flaws in the "adaptationist programme," that of atomizing organisms into traits, which are then individually interpreted as optimally designed for their function.

In the case of a single trait expressed in two environments, the crossing-interaction variance can be conveniently expressed as a genetic correlation (Falconer, 1952; Cockerham, 1963; Via & Lande, 1985). The expressions of one trait in two different environments are interpreted as two traits ("character states") that are genetically correlated. The expression of the trait in the two environments is assumed to have the same genetic basis when the genetic correlation equals one (or negative one). When the correlation is between one and negative one, different genes are assumed to be responsible for the expression of the trait in the two different environments. If the genetic correlation of fitness across environments is negative, then distinct genotypes are better adapted to the distinct environments (*i.e.*, there is a genetic "trade-off") and there is potential for the evolution of specialization. In summary, limits on the natural selection of quantitative traits may be attributable to any of four sources: (1) the magnitude of additive genetic variance within a particular environment (V_A), (2) the magnitude of phenotypic variance expressed within that environment (V_P), relative to the magnitude of the additive genetic variance, (3) the genetic correlation between traits, and (4) the magnitude of genetic variation in response to variation in the natural environment.

Estimation of the elements of E and I in nature is difficult. It requires measurement of a relatively large number of pedigreed individuals in several natural environments. Even for a minimal design a large amount of labor is required, in part because of the notoriously large sampling variance of variance components. A further practical difficulty is that experiments to determine realistic fitness measures in the field will, for most plant species, result in a large number of missing data, due to mortality or reproductive failure.

The problem of estimating variance components based on the resulting unbalanced data has no simple solution in the classical approach using the method of moments (Hartley, 1967). Moreover, that approach provides no method of directly estimating the across-environment covariance nor for testing certain hypotheses about variance components. Maximum-likelihood methods (Shaw, 1987) not only provide a straightforward way of estimating variance components with unbalanced data, but also allow for the statistical comparison of genetic variance components (Klein, 1974; Shaw, 1991). Regardless of the statistical method of analysis, relatively large sample sizes are required in order to reliably estimate variance components (Mitchell-Olds & Rutledge, 1986; Shaw, 1987, 1991). We will illustrate the use of maximum-likelihood in estimating genetic variance components and genetic correlations on data from a field experiment.

A CASE STUDY: *ANTHOXANTHUM ODORATUM* IN THE CALIFORNIA COASTAL PRAIRIE

Coastal grasslands at the Sea Ranch, California (180 km N of San Francisco), were invaded by the introduced perennial grass *Anthoxanthum odoratum* L. within 20 years after cessation of grazing (Peart & Foin, 1985; Foin & Hektner, 1986). Rainfall is extremely seasonal in these grasslands with 94% of the annual precipitation of 1030 mm occurring between October and April (U.S. Environmental Data and Information Service, 1951-1975). The length of the growing season varies in the grasslands, depending on the local drainage conditions. In well-drained xeric sites the growing season for *A. odoratum* may be 4 to 6 wk shorter than in mesic swales (Fig. 1). As a result, the species composition, biomass and light regime differ significantly between xeric and mesic sites (Fig. 2; Table 1).

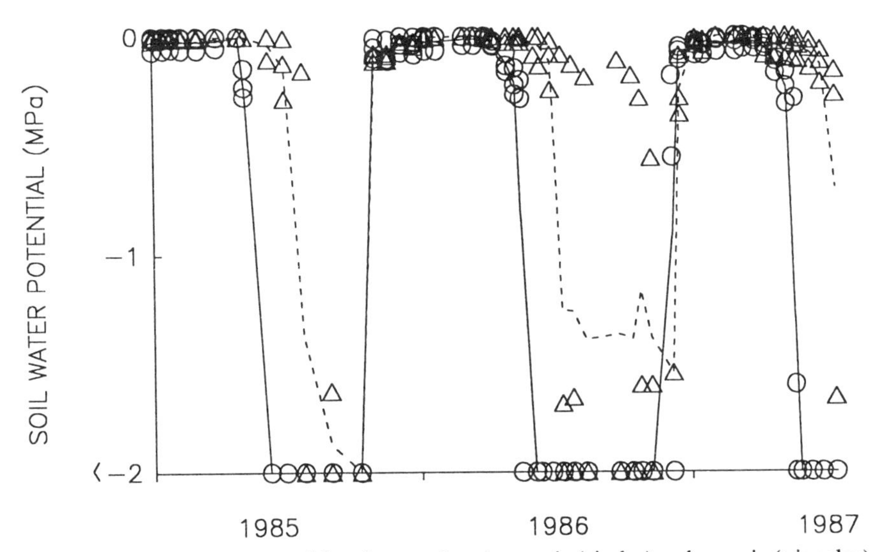

FIGURE 1. Soil water potential at three stations in a xeric (circles) and a mesic (triangles) grassland site at Sea ranch, California, and their averages (solid and dashed line, respectively). Measurements were made at 35 cm depth (data from Platenkamp, 1990).

A. odoratum rapidly invaded sites with widely differing soil moisture regimes. A series of reciprocal transplant experiments between mesic and xeric sites revealed remarkable phenotypic plasticity in some traits and significant genetic differentiation between mesic and xeric populations in other traits, while some traits showed both genetic differentiation and phenotypic plasticity (Platenkamp & Foin, 1990; Platenkamp, 1990, 1991). The patterns of genetic differ-

TABLE 1. Characteristics of xeric and mesic grassland sites at the Sea Ranch, California (data from Platenkamp & Foin, 1990).

	Xeric Site	*Mesic Site*
Mean standing crop (± SE) (Kg/0.25 m^2)	0.21 ± 0.01	0.34 ± 0.03
Penetration of PAR (%)[1]	9.0 (0.1 - 70.4)	0.1 (0 - 1.0)

[1] Mean and range of photosynthetically active radiation (PAR) at 4.5 cm above the soil surface.

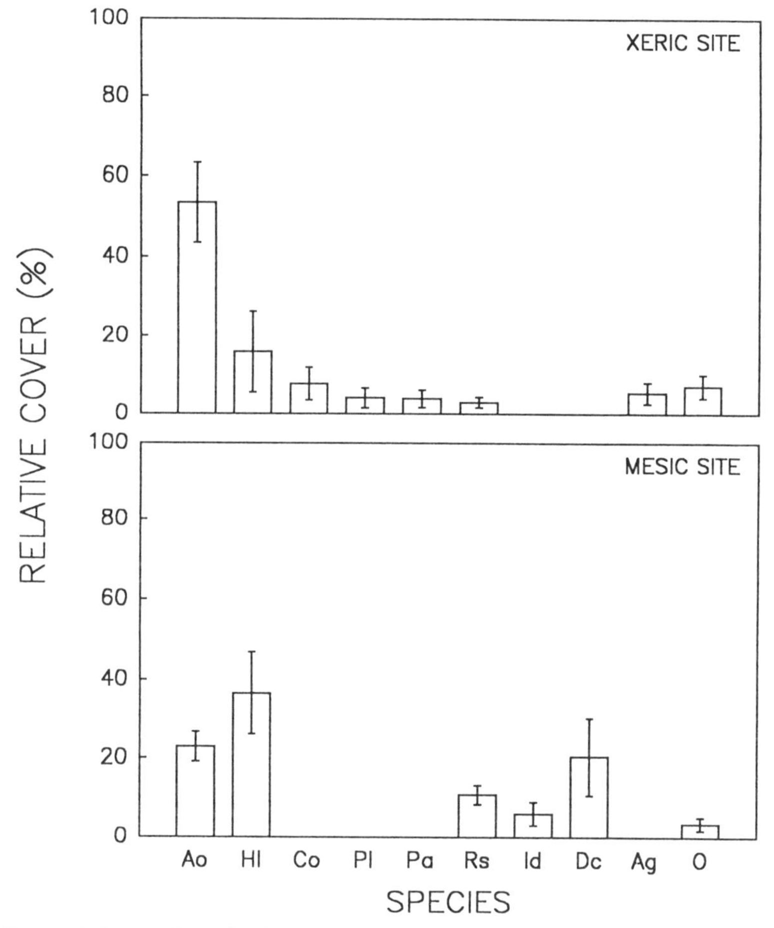

FIGURE 2. Cover of grassland plant species at two sites with different soil moisture regimes at the Sea Ranch, California, means (± SE) of five estimates obtained with a "parallax frame." Ao = *Anthoxanthum odoratum*, Hl = *Holcus lanatus*, Co = *Convolvulus occidentalis*, Pl = *Plantago lanceolata*, Pa = *Pteridium aquilinum*, Ru = *Rubus ursinus*, Id = *Iris douglasiana*, Dc = *Deschampsia cespitosa* ssp. *holciformis*, Ag = annual grasses, O = other species (data from Platenkamp, 1990).

entiation were largely consistent with the hypothesis of divergent selection in the two habitats (Platenkamp, 1990). We will demonstrate this here for first year reproductive output.

Evidence For Prior Adaptive Differentiation

A transplant experiment with cloned tillers showed significant genetic differences in growth and reproductive timing between populations, but not in mortality. Mortality was strongly seasonal, and the level of mortality was site dependent. In the xeric site, levels of mortality were much higher than in the mesic site (Fig. 3). After the first summer of the experiment almost half of the transplants to the

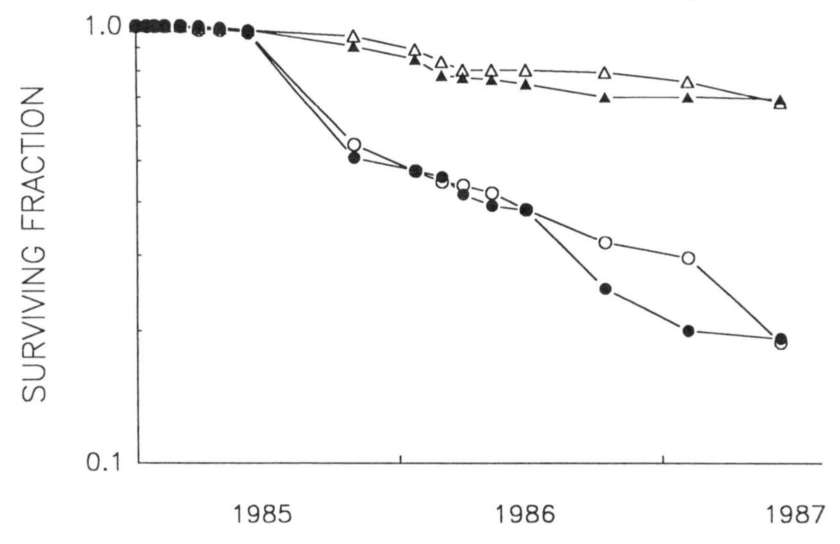

FIGURE 3. Age-dependent survivorship of *Anthoxanthum odoratum* clones transplanted between a xeric (circles) and mesic (triangles) site at Sea Ranch, California. Xeric origin (open symbols): N = 224, mesic origin (closed symbols): N = 240 (data from Platenkamp, 1990).

xeric site had died, while fewer than 10% had died in the mesic site. Genetic differentiation between these two populations was further evidenced in comparisons of several additional characters. The time to flowering within the first and second years was shorter for clones originating from the xeric site, than for clones originating from the mesic site. Size-dependent survival, growth and reproduction also showed significant differences between the populations of origin.

Focussing here on first year reproduction we ask whether the population difference is adaptive. We hypothesized that a relatively high summer mortality would select phenotypes with a relatively high first year reproductive output. This was indeed the case (Fig. 4). Clones originating from the xeric site produced more inflorescences in the first year and also had an earlier average age of first reproduction than clones originating from the mesic site.

The fact that clones that originated from the xeric site had a higher first year reproductive output in both environments than mesic origin clones (Fig. 4) suggests that xeric origin clones are better adapted to both environments, *ceteris paribus*. This could be interpreted as

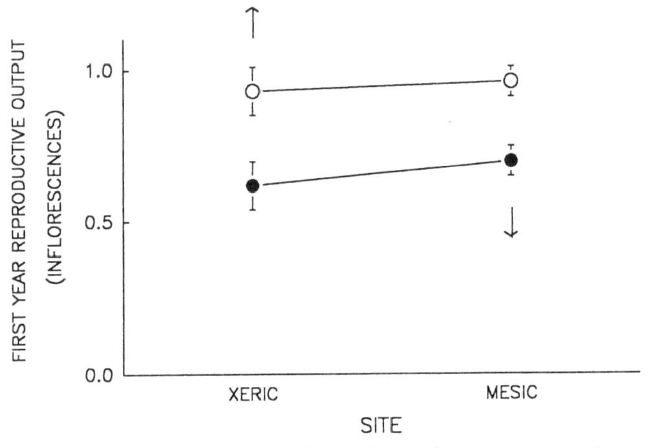

FIGURE 4. Mean reproductive output for *Anthoxanthum odoratum* clones transplanted between a xeric and a mesic site at Sea Ranch, California. Filled symbols = mesic population, open symbols xeric population. The arrows indicate the direction of selection (data from Platenkamp & Shaw, 1992).

absence of genotype-environment "crossing" interaction. However, if total lifetime fitness is associated with high first year reproductive output in the xeric site, but with low first year reproductive output in the mesic site, then the identical ranks represent evidence *for* local adaptation. In fact, the clonal correlation between first year reproductive output and total three year reproductive output (a measure of individual fitness) was positive for both populations in the xeric site ($r = 0.86$ and 0.82 for the xeric and mesic populations, respectively), but negative in the mesic site for the mesic population ($r = -0.70$) and inestimable for the xeric population (but the covariance was negative). This indicates that selection on first year reproductive output may be in opposite directions in the two sites. In the mesic site, a relatively high first year reproductive output is selected against, presumably because a cost of reproduction outweighs the risk of early mortality in this site (Platenkamp, 1990; Platenkamp & Shaw, 1992). Thus, because selection on first year fecundity favors opposite ends of the phenotype distribution in the two environments, the naive conclusion that the xeric population is better adapted to both sites is erroneous. On this evidence, each population is better adapted to its site of origin.

Limits to Future Differentiation

A quantitative genetic analysis of clonal variation in the transplant data was used to infer possible limits to evolution of growth and reproductive allocation. Regarding the potential for future evolution by natural selection we can make the following predictions: (1) the rate of future evolution within each population depends on the magnitude of the additive genetic variance relative to the total phenotypic variance, (2) negative correlations between fitness components, such as between first and second year reproductive output, may limit the range of potential combinations of these characters, and (3) further differentiation of the two populations as a result of local adaptation is promoted by "crossing" genotype-environment interaction as evidenced by across-environment genetic correlations less than unity. We used the variance among clones as an upper bound for additive genetic variance, because clonal variance may contain dominance variance, epistatic variance and maternal genetic and maternal environmental variance, in addition to additive genetic variance.

CLONAL AND ENVIRONMENTAL VARIATION

Again, we use first year reproductive output as an example. Although the environmental variation does not differ between sites or populations of origin for this character, there appears to be a difference in the expression of clonal variation in the two environments (Table 2). For both populations of origin significant clonal variation is found in the xeric site but not in the mesic site. Heritabilities in the mesic site are very low, indicating that any evolution of first year reproductive output would be very slow at this site. Although heritabilities are moderate in the xeric site, these only represent an upper bound for the genetic variation that can provide raw material for selection. In particular, carry-over or "maternal environmental" effects on clones have been previously documented in plants (Hume &

TABLE 2. Restricted maximum-likelihood (REML) estimates of clonal (V_C) and environmental (V_E) variance components (x 1000), and broad sense heritability (H^2) for first year reproductive output (data from Platenkamp & Shaw, 1992).

		SITE					
		XERIC			*MESIC*		
		V_C	V_E	H^1	V_C	V_E	H^1
POPULATION	*XERIC*	5.7[2]	19.1	0.23	(-1.2)	21.1	0
	MESIC	7.4[1]	20.5	0.27	1.1	22.0	0.05

[1] $p < 0.05$, [2] $p < 0.01$

Cavers, 1981; Antonovics, *et al.*, 1987) and these may have contributed to the clonal variance.

CLONAL CORRELATIONS BETWEEN TRAITS

Relatively high expenditures to reproduction in the first year may result in less growth and reproduction in subsequent years, if there is a cost of reproduction. Moreover, if such a trade-off has a genetic basis, it can impede response to selection. To evaluate costs of reproduction we examined the correlations between reproductive output in the first year and growth and reproductive output in subsequent years. Here, we only consider the clones transplanted to their respective sites of origin.

In the xeric population the clonal correlation between first year reproductive output and second year growth is positive (Table 3). The correlation between first and second year reproductive output is inestimable, but the covariance is positive. Thus, no evidence for a cost to reproduction was found in the xeric site. For the mesic population the clonal correlations between first year reproductive

TABLE 3. Estimates of genetic correlations for the xeric population transplanted into the xeric site. The sign of the covariance is given in brackets when correlations were inestimable. RO = reproductive output (data from Platenkamp & Shaw, 1992).

	GROWTH 1986	*RO 1985*	*RO 1986*
GROWTH 1985	-0.07	0.79[1]	(+)
GROWTH 1986		0.31	(+)
RO 1985			(+)

[1] $p = 0.06$

output and growth and reproduction in the following years were negative (Table 4). There appears to be a cost to first year reproduction in this site. Such a cost might be incurred if production of inflorescences is costly in the mesic site because low light levels in the vegetation limit carbon acquisition. The negative correlation represents a limit on evolution of first year reproductive output, because it is likely to result in evolution of a joint response with lower reproductive expenditures in the first year and higher ones in the second year.

TABLE 4. Estimates of genetic correlations for the mesic population transplanted into the mesic site. RO = reproductive output. Due to sampling error, correlations can be larger than 1, or smaller than -1. Significance levels are for tests of the null hypothesis that the covariance equals 0 (data from Platenkamp & Shaw, 1992).

	GROWTH 1986	*GROWTH 1987*	*RO 1985*	*RO 1986*	*RO 1987*
GROWTH 1985	1.38	1.91[1]	-0.33	1.97[1]	2.18[1]
GROWTH 1986		1.05	-1.06	1.04	1.05
GROWTH 1987			-0.07	0.73	0.87
RO 1985				-1.00	-0.19
RO 1986					0.74

[1] $p < 0.05$

GENOTYPE-ENVIRONMENT INTERACTION

Clonal variance for first year reproductive output tended to be larger in the xeric site than in the mesic site (Table 2), although the difference was not statistically significant (Platenkamp & Shaw, 1992). Differences in the magnitude of genetic variance between sites represent a source of genotype-environment interaction variance. However, in relation to the evolution of genetic specialization (*e.g.*, population differentiation), the more relevant component of interaction variance is the variation due to a difference in the ranking of clones in the two environments. This variation was evaluated by calculating the across-environment correlation. For the xeric population this correlation could not be calculated, because the clonal variance component for the mesic site was negative (Table 2). However, the covariance was positive. For the mesic population an across-environment genetic correlation close to zero was found (- 0.06). In addition, the covariances were not significantly different from zero. When all clones were considered as part of one combined population, the correlation for first year reproductive output was estimated to be one. Because selection on this trait has been shown to operate in opposite directions in the two environments, this positive correlation suggests an underlying trade-off for this trait, indicating the possibility of the evolution of divergence.

In summary, the estimation of clonal components of variance and covariance illustrates evolutionary limitations in the mesic population. In that population, response to selection on first year reproductive output will be slow, because the heritability is relatively low. Moreover, the pattern of clonal correlations indicates that reproductive output in the first and second years cannot both be increased. This pattern, in association with the genetic correlation between these components of fitness and lifetime fitness, suggests that reproductive expenditure is in the process of shifting from the first to the later years in this population. Concerning the process of differentiation between the populations, we infer that divergence in first year reproductive output could proceed, given the divergent selection between populations for this component of the life history. Our example illustrates the hazard of the naive assumption that selection necessarily favors increases in fecundity at each stage of the life cycle.

QUANTITATIVE VARIATION IN NATURAL PLANT POPULATIONS

Quantitative genetic variation is present in the majority of organisms where it has been studied (Barton & Turelli, 1989), but little is known about narrow sense heritabilities in heterogeneous natural environments, or about genotype-environment interactions in nature. Information concerning both of these is necessary for predicting evolution by natural selection in the wild (Mitchell-Olds & Rutledge, 1986). A few estimates of heritabilities and genetic correlations are available in the literature for natural plant populations. We will review most of them here in an attempt to assess the generality of our findings for *A. odoratum*.

Heritability estimates from 16 plant studies are summarized in Table 5. Seed characters were excluded because these are inherited through multiple pathways and are expressed in multiple environments (see Platenkamp & Shaw, 1993). Table 5 shows a relatively wide range of heritability values. This is perhaps not surprising, considering the different breeding systems, population histories, ranges of characters, experimental procedures and statistical methods represented in these studies. In most studies, some characters had heritabilities that were not significantly different from zero. In several studies this was true for the majority of estimates (*e.g.*, Mitchell-Olds, 1986; Schwaegerle & Levin, 1991; Shaw & Billington, 1991; Kelly, 1993; Shaw & Platenkamp, 1993). These results suggest that response to selection may often be rather slow, given the preponderance of small heritabilities (0.1), in spite of earlier claims suggesting the contrary (Antonovics, 1976). Nonetheless, low narrow sense heritabilities were not universal in plant studies.

We found only three studies that reported narrow sense heritabilities based on field data (Mitchell-Olds, 1986, Schwaegerle & Levin, 1991; Mazer and Schick, 1991). Schwaegerle & Levin (1991) found that five of the 14 heritability estimates in *Phlox drummondii* were significantly larger than zero ($h^2 = 0.06 - 0.15$), while additive genetic variance in six of the remaining nine traits was estimated to be equal to zero. Mitchell-Olds (1986) found that in one population of *Impatiens capensis* none of 13 heritability estimates were significantly larger than zero, while in the other population two estimates were significantly larger than zero ($h^2 = 0.29$ and 0.32, respectively). Nine of the 26 estimates of heritability in his study were equal to zero. A reanalysis of four non-seed traits changed the median estimates in Mitchell-Olds (1986) only slightly, to 0 and 0.11, respectively for the two populations (Mitchell-Olds & Bergelson, 1990).

In a study on reproductive traits in *Raphanus sativus* Mazer & Schick (1991) found narrow sense heritability estimates that were relatively high, considering that they were obtained in a field study (median $h^2 = 0.35$). The paternal variance component was higher than the maternal variance component in five out of nine cases (their Fig. 4). This is surprising since maternal variance components are usually larger than paternal variance components in plants, due to the preponderance of maternal effects and, in any case, are expected to be at least as great as paternal components.

Table 5 shows that broad sense estimates from natural field environments tended to be smaller than broad sense estimates obtained in the greenhouse, as expected. This simply may reflect the larger environmental variation in the field. Broad sense estimates can be used to estimate an upper bound to narrow sense heritability. That is, if they are low, we may assume that narrow sense heritabilities are

TABLE 5. Median heritability estimates for natural plant populations. Seed characters were excluded. Traits measured on different populations were tabulated separately (population given in brackets). Traits measured in different environments were treated as separate traits. For reciprocal transplants only the estimates for the populations at the "home" site were used (except where indicated). N = the number of traits. F/G = field (F), versus greenhouse, growthchamber or garden (G).

Species	*Median*	*N*	*F/G*	*Source*
Narrow sense estimates:				
Impatiens capensis (Ma)	0.02	13	F	Mitchell-Olds, 1986
Impatiens capensis (Mi)	0.10	13	F	Mitchell-Olds, 1986
Phlox drumondii	0.01	14	F	Schwaegerle & Levin, 1991
Raphanus sativus	0.35	9	F	Mazer & Schick, 1991
Chamaecrista fasciculata	0.01	10	G	Kelley, 1993
Holcus lanatus (I)	0.02	6	G	Shaw & Billington, 1991[2]
Holcus lanatus (T)	0.04	6	G	Shaw & Billington, 1991[2]
Nemophila menziesii	0.07	9	G	Shaw & Platenkamp, 1993
Raphanus raphanistrum	0.04/0.14	9	G	Mazer, 1987[1]
Plantago lanceolata (M)	0.26	14	G	Wolff, 1990
Plantago lanceolata (W)	0.44	13	G	Wolff, 1990
Senecio integrifolius (G)	0.28	30	G	Widén & Andersson, 1993
Senecio integrifolius (B)	0.51	30	G	Widén & Andersson, 1993
Broad sense estimates:				
Anthoxanthum odoratum (X)	0.22	7	F	Platenkamp & Shaw, 1992
Anthoxanthum odoratum (M)	0.07	7	F	Platenkamp & Shaw, 1992
Danthonia spicata (CH)	0.22	12	F	Clay and Antonovics, 1985[3]
Danthonia spicata (CL)	0.22	12	F	Clay and Antonovics, 1985[3]
Plantago lanceolata (B)	0.15	10	F	V. Tienderen & V.D. Toorn, 1991
Plantago lanceolata (H)	0.13	10	F	V. Tienderen & V.D. Toorn, 1991
Plantago lanceolata (J)	0.03	10	F	V. Tienderen & V.D. Toorn, 1991
Salvia lyrata	0.10	8	F	Shaw, 1986
Spartina patens	0.21	22	F	Silander, 1985[4]
Spartina patens	0.29	17	G	Silander, 1985[4]
Heterosperma pinnatum (S)	0.28	9	G	Venable & Búrquez, 1989[5]
Heterosperma pinnatum (T)	0.50	9	G	Venable & Búrquez, 1989[5]
Heterosperma pinnatum (G)	0.43	9	G	Venable & Búrquez, 1989[5]
Heterosperma pinnatum (H)	0.43	9	G	Venable & Búrquez, 1989[5]
Heterosperma pinnatum (Z)	0.48	9	G	Venable & Búrquez, 1989[5]
Heterosperma pinnatum (M)	0.46	9	G	Venable & Búrquez, 1989[5]
Poa annua (G)	0.23	9	G	Till-Bottraud, *et al.*, 1990
Poa annua (R)	0.87	11	G	Till-Bottraud, *et al.*, 1990

[1] Data from the two blocks of the reciprocal factorial breeding design were analyzed separately; [2] Unconstrained estimates from Table 1A; [3] CH and CL refer to progeny of chasmogamous and cleistogamous flowers on plants from one population; [4] Clones of (sub)populations were pooled; [5] Only life-history, size and shape traits; complete selfing was assumed.

low; but if they are high, we have learned little about the causes of variation among genotypes or families. Estimates from field studies are most useful here, because they estimate relevant environmental variation, while greenhouse studies tend to be of limited value for estimating evolutionary rates. Some studies that used broad sense estimates suggest that for many traits little response to selection can be expected in the wild (Shaw, 1986; Van Tienderen & van der Toorn, 1991; Platenkamp & Shaw, 1992).

There is remarkable variation among estimates obtained from different populations, which suggests that the potential to respond to selection differentials is not constant throughout a species' range (Table 5). However, only Mitchell-Olds (1986), Shaw & Billington (1991) and Platenkamp & Shaw (1992) tested the differences in the genetic variance components between populations. Mitchell-Olds (1986) compared heritabilities for three traits between populations using randomization tests and found that one was marginally significant, while two were non-significant. Shaw & Billington (1991) and Platenkamp & Shaw (1992) used maximum-likelihood tests to compare variance components between populations (Shaw, 1991). For *A. odoratum* none of the tests were significant (Platenkamp & Shaw, 1992), while for *Holcus lanatus* a significant population difference was found for one character under very specific assumptions (Shaw & Billington, 1991).

Some heritability estimates in Table 5 must be interpreted with caution. For instance, the estimates for *Heterosperma pinnatum* assume complete selfing, although this species is only a partial selfer, so the heritabilities have probably been overestimated (Venable & Búrquez, 1989). In contrast, Mitchell-Olds (1986) corrected heritability estimates for the degree of inbreeding in the population using estimates from electrophoretic data. Mitchell-Olds & Rutledge (1986) provide a thorough discussion of the assumptions underlying quantitative genetic theory and their consequences for the estimation of components of variance and covariance in natural plant populations.

Few studies report genetic correlations between traits measured in the field. Mitchell-Olds (1986) provides narrow sense genetic correlations for a *Impatiens capensis* population, measured in the field. The correlations between four non-seed characters ranged between 0.69 and 1.53. Schwaegerle & Levin (1991) estimated genetic correlations between six traits related to female reproductive success in a *Phlox drummondii* population. Estimates ranged from -1.07 to 1.06, while the covariances in both extreme cases were significantly different from zero. However, the significantly negative correlations involved the shoot/root ratio, which was negatively correlated to fitness. So, in this population no negative correlations between fitness components were found. Shaw (1986) estimated broad sense genetic correlations between leaf number and leaf length for *Salvia lyrata* from field data at two densities and two dates. Three of the correlations were significantly greater than zero, while one estimate was negative ($r = -0.64$) but not significantly different from zero. These studies do not provide evidence that negative genetic correlations between traits generally constitute a limit on evolution by natural selection of plants in the field.

Genetic correlations for the expression of characters between stands with different densities in the field were estimated for leaf number and leaf length in *S. lyrata* by Shaw (1986). Variance component correlations for leaf length ranged from 0.32 to 0.89, while correlations for leaf number ranged from 0.33 to 1.08. We may conclude that some degree of divergence may evolve when these correlations differ significantly from one (Gillespie & Turelli, 1989). Nonetheless, no evidence was found for across-environment genetic trade-offs because all correlations were larger than zero (assuming selection in the same direction in both environments). Such trade-offs would indicate that, on average, different families have relatively higher performance in one environment than in the other. Van Tienderen & van der Toorn (1991) calculated clonal across-environment correlations for three reciprocally transplanted *Plantago lanceolata* populations. In three pair-wise comparisons for ten traits the estimates for three traits were negative, but they were not significantly different from zero. Fifteen of the positive correlations were significantly larger than zero. Again, in this case there was no evidence for a genetic trade-off with different clones being favored by selection in different environments.

Stratton (1992) calculated broad sense genetic correlations for size and fecundity in *Erigeron annuus* across three competitive regimes. Estimates for size tended to be weakly negative, but for fecundity they were positive. Shaw & Platenkamp (1993) estimated narrow sense genetic correlations for dry weight of *Nemophila menziesii* between different competitive regimes in a greenhouse. The correlations ranged from 0.13 to 1.00, with none of the correlations being significantly smaller than one. So far, there appears to be little published evidence for genetic trade-offs in performance (fitness) across environments for natural plant populations (but see Khan & Bradshaw, 1976, for a crop example).

Only a small number of studies have so far attempted to estimate the quantitative genetic variation in natural populations of plants in different environments. Most of these studies were done in a greenhouse or garden environment and most were concerned with broad sense measures of variation. We have focussed here on natural plant populations in natural environments. The available evidence suggests that narrow sense heritabilities expressed under natural condition often do not exceed 0.1. Heritabilities of this magnitude could, in principle, support a gradual response to selection, when considered over several generations. But statistical precision of estimates in this range is poor, and power to test them is weak. For few of these estimates is it possible to reject the null hypothesis that heritability is zero, corresponding to prediction of no selection response. Similarly, few examples of within-population genetic variation in response to the environment have been documented in natural populations. We found no evidence for adverse genetic correlations within populations as a limit on the direction of future evolution.

A hundred years after the birth of Göte Turesson, the father of the study of ecogeographical races, we have still very much to learn about what limits the evolution of locally adapted populations. A combination of experimental demography and quantitative genetic analysis can help us gain insight in these limits on a relatively short time scale. Over the long term the supply of genetic variation may be a more relevant area of study (Bradshaw, 1991). One such attempt is the recently initiated study of mutational variance in *Arabidopsis thaliana* by J. Willis and M. Lynch (pers. commun.).

ACKNOWLEDGMENTS

We thank Stefan Andersson, John Bishop, Joe Felsenstein, and Robert Podolsky for their constructive comments on the manuscript.

Literature Cited

Antonovics, J. 1976. The nature of limits to natural selection. *Ann. Missouri Bot. Gard.* 63:224-247.

Antonovics, J., K. Clay, and J. Schmitt. 1987. The measurement of small-scale environmental heterogeneity using clonal transplants of *Anthoxanthum odoratum* and *Danthonia spicata*. *Oecologia* 71:601-607.

Antonovics, J., N. C. Ellstrand and R. N. Brandon. 1988. Genetic variation and environmental variation: Expectations and experiments. Pages 275-303 *in* L. D. Gottlieb & S. K. Jain, eds., *Plant Evolutionary Biology*. Chapman & Hall, New York.

Antonovics, J. and R. B. Primack. 1982. Experimental ecology and genetics in *Plantago*. VI. The demography of seedling transplants of *P. lanceolata*. *Jour. Ecol.* 70:55-75.

Antonovics, J., and P. H. van Tienderen. 1991. Ontoecogenophyloconstraints? The chaos of constraint terminology. *Trends Ecol. Evol.* 6:166-167.

Barton, N. H. and M. Turelli. 1989. Evolutionary quantitative genetics: How little do we know? *Annu. Rev. Genetics* 23: 337-370.

Bradshaw, A. D. 1991. Genostasis and the limits to evolution. *Philosophical Trans. Royal Soc. London*, ser. B, 333:289-305.

Clausen, J., D. D. Keck and W. M. Hiesey. 1940. Experimental studies on the nature of species. I. The effect of varied environments on western American plants. *Carnegie Inst. Washington, Publ.* 520. 450 pp.

Clausen, J., D. D. Keck, and W. M. Hiesey. 1948. Experimental studies on the nature of species. III. Environmental responses of climatic races of *Achillea*. *Carnegie Inst. Washington, Publ.* 581. 129 pp.

Clay, K. and J. Antonovics. 1985. Quantitative variation of progeny from chasmogamous and cleistogamous flowers in the grass *Danthonia spicata*. *Evolution* 39:335-348.

Clements, F. E., and G. W. Goldsmith. 1924. The Phytometer Method in Ecology. *Carnegie Inst. Washington, Publ.* 356.

Cockerham, C. C. 1963. Estimation of genetic variances. Pages 53-94 *in* W. D. Hanson & H. F. Robinson, eds., *Statistical Genetics and*

Plant Breeding. Natl Res. Council, Publ. 982. Washington, DC, USA.

Davies, M. S. and R. W. Snaydon. 1976. Rapid population differentiation in a mosaic environment. III. Measures of selection pressures. *Heredity* 36:59-66.

Falconer, D. S. 1952. The problem of environment and selection. *American Nat.* 86:293-298.

Falconer, D. S. 1989. *Introduction to Quantitative Genetics*, 3rd ed. Longman, NewYork

Foin, T. C., and M. M. Hektner. 1986. Secondary succession and the fate of native species in a California coastal prairie community. *Madroño* 33:189-206.

Fowler, N. L., and J. Antonovics. 1981. Small-scale variability in the demography of transplants of two herbaceous species. *Ecology* 62:1450-1457.

Gillespie, J. H., and M. Turelli. 1989. Genotype-environment interactions and the maintenance of polygenic variation. *Genetics* 121:129-138.

Gould, S. J., and R. C. Lewontin. 1979. The spandrels of San Marco and the Panglossian paradigm. A critique of the adaptationist programme. *Proc. Roy. Soc. B* 205:581-598.

Gurevitch, J. 1992a. Sources of variation in leaf shape among two populations of *Achillea lanulosa*. *Genetics* 130:385-394.

Gurevitch, J. 1992b. Differences in photosynthetic rate in populations of *Achillea lanulosa* from two altitudes. *Functional Ecol.* 6:568-574.

Hamrick, J. L. 1982. Plant population genetics and evolution. *American Jour. Bot.* 69:1685-1693.

Harper, J. L. 1977. *Population Biology of Plants*. Academic Press, London, UK.

Harper, J. L., and J. White. 1974. The demography of plants. *Annu. Rev. Ecol. Syst.* 5:419-463.

Hartley, H. O. 1967. Expectations, variances and covariances of ANOVA mean squares by "synthesis." *Biometrics* 23:105-114.

Heslop-Harrison, J. 1964. Forty years of genecology. *Adv. Ecol. Res.* 2:159-247.

Hume, L., and P. B. Cavers. 1981. A methodological problem in genecology. Seeds versus clones as source material for uniform gardens. *Canadian Jour. Bot.* 59:763-768.

Jain, S. K., and A. D. Bradshaw. 1966. Evolutionary divergence among adjacent plant populations. I. Evidence and its theoretical analysis. *Heredity* 21:407-441.

Khan, M. A., and A. D. Bradshaw. 1976. Adaptation to heterogeneous environments. II. Phenotypic plasticity in response to spacing in *Linum. Australian Jour. Agricul. Res.* 27:519-531.

Kelly, C. A. 1993. Quantitative genetics of size and phenology of life-history traits in *Chamaecrista fasciculata. Evolution* 47:99-97.

Kelley, S. E., and K. Clay. 1987. Interspecific competitive interactions and the maintenance of genotypic variation within two perennial grasses. *Evolution* 41:92-103.

Klein, T. W. 1974. Heritability and genetic correlation: Statistical power, population comparisons, and sample size. *Behav. Genetics* 4:92-103.

Lande, R. 1979. Quantitative genetic analysis of multivariate evolution, applied to brain: body size allometry. *Evolution* 33:402-416.

Langlet, O. 1971. Two hundred years genecology. *Taxon* 20:653-722.

Loeschcke, V. 1987. Introduction: Genetic constraints on adaptive evolution and the evolution of genetic constraints. Pages 1-2 *in* V. Loeschcke, ed., *Genetic Constraints on Adaptive Evolution.* Springer-Verlag, Berlin, Heidelberg, Germany.

Maynard Smith, J. 1983. The genetics of stasis and punctuation. *Annu. Rev. Genetics* 17:11-25.

Mazer, S. J. 1987. The quantitative genetics of life history and fitness components in *Raphanus raphanistrum* L. (Brassicaceae): Ecological and evolutionary consequences of seed-weight variation. *American Nat.* 130:891-914.

Mazer, S. J., and C. T. Schick. 1991. Constancy of population parameters for life-history and floral traits in *Raphanus sativus* L. II. Effects of planting density on phenotype and heritability estimates. *Evolution* 45:1888-1907.

Mitchell-Olds, T. 1986. Quantitative genetics of survival and growth in *Impatiens capensis. Evolution* 40:107-116.

Mitchell-Olds, T., and J. Bergelson. 1990. Statistical genetics of an annual plant, *Impatiens capensis.* I. Genetic basis of quantitative variation. *Genetics* 124:407-415.

Mitchell-Olds, T., and J. J. Rutledge. 1986. Quantitative genetics in natural plant populations: A review of the theory. *American Nat.* 127:379-402.

Peart, D. R., and T. C. Foin. 1985. Analysis and prediction of population and community change: A grassland case study. Pages 313-339 *in* J. White, ed., *The Population Structure of Vegetation*. Handbook of Vegetation Science. Junk, Dordrecht, The Netherlands.

Perrin, N., and J. Travis. 1992. On the use of constraints in evolutionary biology and some allergic reactions to them. *Functional Ecol.* 6:361-363.

Platenkamp, G. A. J. 1990. Phenotypic plasticity and genetic differentiation in the demography of the grass *Anthoxanthum odoratum*. *Jour. Ecol.* 78:772-788.

Platenkamp, G. A. J. 1991. Phenotypic plasticity and population differentiation in seeds and seedlings of the grass *Anthoxanthum odoratum*. *Oecologia* 88:515-520.

Platenkamp, G. A. J., and T. C. Foin. 1990. Ecological and evolutionary importance of neighbors in the grass *Anthoxanthum odoratum*. *Oecologia* 83:201-208.

Platenkamp, G. A. J., and R. G. Shaw. 1992. Environmental and genetic constraints on adaptive population differentiation in *Anthoxanthum odoratum*. *Evolution* 46:341-352.

Platenkamp, G. A. J., and R. G. Shaw. 1993. Environmental and genetic maternal effects on seed characters in *Nemophila menziesii*. *Evolution* 47:540-555.

Rawson, P. D., and T. J. Hilbish. 1991. Genotype-environment interaction for juvenile growth in the hard clam *Mercenaria mercenaria* (L.). *Evolution* 45:1924-1935.

Reznick, D. N., E. Perry and J. Travis. 1986. Measuring the cost of reproduction: A comment on papers by Bell. *Evolution* 40:1338-1344.

Schmidt, K. P., and D. A. Levin. 1985. The comparative demography of reciprocally sown populations of *Phlox drummondii* Hook. I. Survivorships, fecundities, and finite rates of increase. *Evolution* 39:396-404.

Schwaegerle, K. E., and D. A. Levin. 1991. Quantitative genetics of fitness traits in wild populations of *Phlox*. *Evolution* 45: 169-177.

Shaw, R. G. 1986. Response to density in a wild population of the perennial herb *Salvia lyrata*: Variation among families. *Evolution* 40:492-505

Shaw, R. G. 1987. Maximum-likelihood approaches applied to quantitative genetics of natural populations. *Evolution* 41:812-825.

Shaw, R. G. 1991. The comparison of quantitative genetic parameters between populations. *Evolution* 45:143-151.

Shaw, R. G., and H. L. Billington. 1991. Comparison of variance components between two populations of *Holcus lanatus*: A reanalysis. *Evolution* 45:1287-1289.

Shaw, R. G., and G. A. J. Platenkamp. 1993. Quantitative genetics of response to competitors in *Nemophila menziesii*: A greenhouse study. *Evolution* 47:801-812.

Silander, J. A. 1985. The genetic basis of the ecological amplitude of *Spartina patens*. II. Variance and correlation analysis. *Evolution* 39:1034-1052.

Snaydon, R. W. 1970. Rapid population differentiation in a mosaic environment. I. The response of *Anthoxanthum odoratum* populations to soils. *Evolution* 24:257-269.

Stratton, D. A. 1992. Life-cycle components of selection in *Erigeron annuus*: II. Genetic variation. *Evolution* 46:107-120.

Till-Bottraud, I., L. Wu, and J. Harding. 1990. Rapid evolution of life history traits in populations of *Poa annua* L. *Jour. Evol. Biol.* 3:205-224.

Turesson, G. 1922. The genotypical response of the plant species to the habitat. *Hereditas* 3:211-350.

Turesson, G. 1925. The plant species in relation to habitat and climate. *Hereditas* 6:147-236.

U.S. Environmental Data and Information Service. 1951-1975. *Climatological Data; California Section.* U.S. Environmental Data Service, Asheville, North Carolina.

Van Tienderen, P. H., and J. van der Toorn. 1991. Genetic differentiation between populations of *Plantago lanceolata*. II. Phenotypic selection in a transplant experiment in three contrasting habitats. *Jour. Ecol.* 79:43-59.

Venable, D. L., and A. Búrquez M. 1989. Quantitative genetics of size, shape, life-history, and fruit characteristics of the seed-heteromorphic composite *Heterosperma pinnatum*. I. Variation within and among populations. *Evolution* 33:881-895.

Via, S. 1991. The genetic structure of host plant adaptation in a spatial

patchwork: Demographic variability among reciprocally transplanted pea aphid clones. *Evolution* 45:827-852.
Via, S., and R. Lande. 1985. Genotype-environment interaction and the evolution of phenotypic plasticity. *Evolution* 39:505-522.
Widén, B., and S. Andersson 1993. Quantitative genetics of life-history and morphology in a rare plant, *Senecio integrifolius. Heredity* 70:503-514.
Wolff, K. 1990. Genetic analysis of ecologically relevant morphological variability in *Plantago lanceolata* L. 5. Diallel analysis of two natural populations. *Theor. Appl. Genet.* 79:481-488.
Wu, L., A. D. Bradshaw, and D. A. Thurman 1975. The potential for evolution of heavy metal tolerance in plants. III. The rapid evolution of copper tolerance in *Agrostis stolonifera. Heredity* 34:165-187.

Floral Morphology and the Effects of Crossing Distance in *Scleranthus*

Linus Svensson
Department of Systematic Botany
University of Lund, Ö. Vallg. 18–20
S-223 61 Lund, Sweden

Floral character variation was studied in the highly inbreeding *Scleranthus annuus*, a tetraploid annual weed in the family Caryophyllaceae. Estimates of the distribution of variation within and between plants and populations from natural populations using several sets of floral characters, showed that c. 50% of the variance was among populations and plants, *i.e.*, available for selection. In addition to metric variation in the sepals and styles, the flowers of *S. annuus* have a variable number of fertile stamens per flower.

Artificial crosses were produced using parents separated by different spatial distances. The result of crosses in two populations showed that stamen number in *S. annuus* is dependent upon crossing distance. Short to intermediate crossing distances resulted in progeny having flowers with the highest increase in the number of fertile stamens relative to progeny produced by self-pollination. The response to crossing was not random among stamen-positions in the flower; certain positions had a higher frequency of fully fertile stamens than other positions.

Variation in stamen number in plants from natural populations of *S. annuus* as well as in the flowers of progeny from artificial crosses is discussed in terms of a partial breakdown of the genetic mechanism(s) controlling the development of stamens in *S. annuus*.

Species of flowering plants exhibit considerable variation in response to both the environment and genotype. The degree of variation is, however, not the same for all the attributes of the plant. It is well known among systematists and taxonomists that different parts of the plant are to different degrees 'useful' as systematic characters, *i.e.*, different characters vary to different degrees within an individual. Many authors have stressed the differences in stability among different types of characters in plants. Clausen, *et al.* (1940) and Stebbins (1950) have in classical works stressed the differences in variability between vegetative and floral characters of a species. The relative stability of floral characters compared with vegetative characters is part of the reason why the former have been intensively utilized for the phylogeny, biosystematics and taxonomy of flowering plants (*e.g.*, Ornduff, 1969; Barrett, 1989). Floral characters, such as the size and shape of floral parts, have, however, been shown to vary within species in response to the environment (*e.g.*, Meagher, 1988; Andersson, 1989), to single-gene mutations (Hilu, 1983) and to mutations in meristems (Ellstrand, 1983). Variation in floral characters for an individual produced by artificial crossing has also been reported to be influenced by the spatial distance separating the parental plants in the population of origin (Svensson, 1988, 1990, 1991).

This paper comprises several parts that deal with natural and experimentally induced variation in floral characters in an autogamous annual, *Scleranthus annuus*. The first part is a report on the variation in floral parts in a number of natural populations of *S. annuus* in Sweden, showing substantial variation in both the size and shape of the sepals and gynoecium as well as the number of stamens in the flower. It was also found that variation in the fertility among the 10 stamens of the flower was not random. Certain positions in the flower exhibited a more pronounced variation than others.

The second part of the paper deals with induced variation in stamen numbers in offspring produced by artificially crossing parental plants that were separated by various spatial distances. Crosses were performed in two natural populations, one in which *S. annuus* occurs continuously and one with the species occurring in patches. Using the two populations, with different spatial structure, it was possible to show that the spatial separation of parental plants utilized for artificial crosses did influence the number of stamens in the flowers of the offspring. Responses dependent on spatial distance, similar to the ones observed in the androecium were also found when studying the size and shape of the gynoecium as well as of one of the sepals, although the effect of the spatial separation of parents on sepal and gynoecium morphology was not as obvious as in the case of the androecium.

The two parts of this paper will stress the importance of adopting alternative views of a feature. The first part concentrates on phenotypic variation among natural populations, while the second part is on induced variation in progeny from within-populational crosses. Taken together the two parts aim at describing variation in stamen number of *S. annuus* and supplying a suggestion as to why stamen number variation is observed in this species.

SCLERANTHUS

The genus *Scleranthus* is represented in northern Europe by the highly selfing *S. annuus* and the ant-pollinated *S. perennis* (Svensson, 1985). *Scleranthus annuus* (Caryophyllaceae) is a tetraploid (Moore, 1982) annual weed that grows in several types of disturbed habitats from arable fields to (semi-) natural, open, sandy grasslands but not in dense grass swards or other types of closed vegetation.

The plant has branched stems, opposite, narrow leaves and one, to several thousand small (diameter 1–2 mm) flowers in cymes. The flower has a simple perianth of greenish sepals (Sattler, 1973; *cf.* Tucker, 1989) with narrow white margins. Being a member of the family Caryophyllaceae, *S. annuus* has 10 obdiplostemonous positions in the flower where a stamen or staminoid can occur (Fig. 1). The stamen positions are located in two whorls around the two styles. The average number of normally developed stamens in a plant growing in its natural habitat is two, located in the inner of the two whorls of stamens (*cf.* Svensson, 1988, 1990, 1991 for a detailed description of the floral morphology of the species). The flowers exhibit little or no protandry; the stamens dehisce and the stigma is receptive at the time when the flower opens or up to a few hours later, each anther having about 100 pollen grains (Svensson, 1985).

For a study of this kind, a complete description of the breeding system of the species being studied is crucial. For *S. annuus* no quantitative estimate of the selfing rate is available. The species is tetraploid and highly selfing, which makes the use of allozymes for the determination of the selfing rate less attractive. A number of characteristics of *S. annuus* indicate that the species is selfing to a very high degree, *viz.*, (1) the species has very small flowers (the diameter is 1–2 mm), (2) the anthers dehisce and the stigma becomes receptive when the flowers open. There is no pronounced delay between the male and female phases of flowering, (3) the flowers do not produce any nectar or other attractants, (4) petals are lacking and (5) the species is not visited by any flower visitors (several hundred hours have been spent observing permanent plots of *S. annuus* in its natural habitat in several populations, but no flower visitor has ever been seen [Svensson, unpubl. data]), (6) the species has a complete seed set under isolation. All these characters taken together indicate that *S. annuus* is a highly selfing plant species, with a degree of selfing close to 1.

The fruit is a single-seeded nutlet which, at the time of dispersal, is still attached to the firm perianth and is passively dispersed around the seed parent.

METHODS

Morphological Measures

The strict location of each stamen position in relation to the sepals allowed an unambiguous identification of each stamen position (Sterk, 1970; Svensson, 1990). The 10 stamen positions of the flowers of *S. annuus*, denoted STA 1 – STA 10 (*cf.* Fig. 1), may hold either staminoids without pollen sacs or with pollen sacs lacking normally-developed pollen grains or stamens with one or two pollen sacs containing normally-developed pollen grains (see Fig. 2). A way of scoring the organs (hereafter "fertility scores") found in the stamen positions in the flower of *S. annuus* was chosen to mirror the array of organs found, with the highest score given to a completely developed, normal stamen, and score 1 given to the most reduced type of staminoid (in addition, '0' which was given to a stamen position lacking any organ, see Fig. 2 and Svensson, 1988, 1990, 1992). It should be noted that only stamens scored 5 or 6 contain viable pollen grains (*i.e.*, pollen grains stained blue by aniline-lactophenol). The summed scores for all stamen positions of a flower (STAMTOT) was derived, together with the summed scores for stamen positions in the inner whorl (STA 2, STA 4, STA 6, STA 8, STA 10) "STAIN" and for the positions in the outer whorl (STA 1, STA 3, STA 5, STA 7, STA 9) "STAOUT."

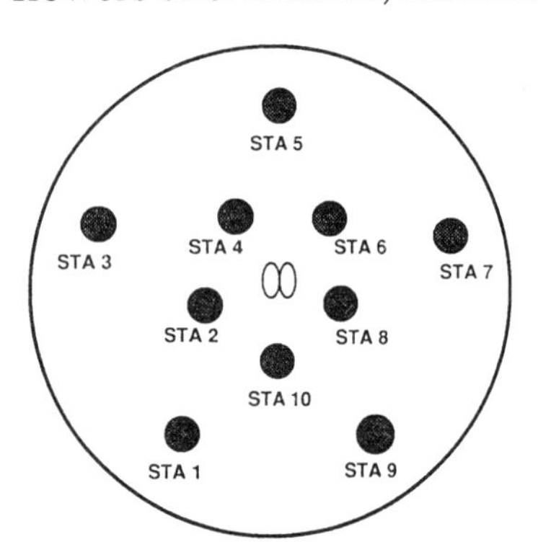

FIGURE 1. Diagrammatic representation of the 10 stamen positions in the flowers of *Scleranthus annuus*. The distance between the two whorls is exaggerated. Two styles are located in the centre.

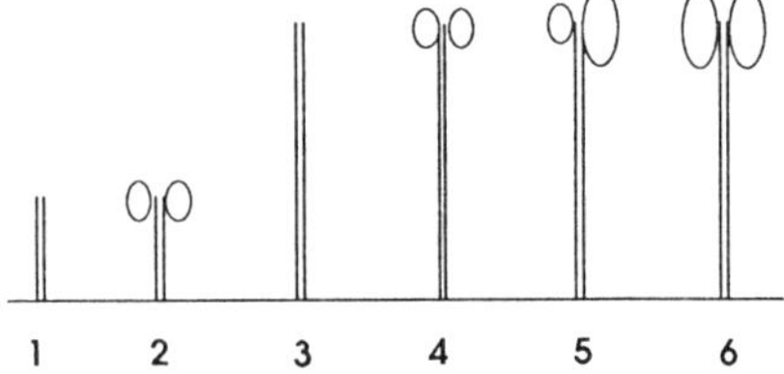

FIGURE 2. The six types of staminoids and stamens in the flowers of *Scleranthus annuus*. Staminoids classified as 1-4 are sterile. Stamens classified as '5' contain one pollen sac with fertile pollen, while stamens in class '6' have two pollen sacs with fertile pollen. A stamen position without any structure was scored as '0.'

For the study of the morphological variation in sepals and gynoecia, I transferred the pistil and the lowermost sepal from each of 10 randomly selected flowers per plant to 10 microscope slides. Style length (STYLELENG) and stigma lengths (STIGMALENG) were measured, together with several sepal characters as shown in Fig. 3, using a microscope, a camera lucida and a digitizer connected to a microcomputer.

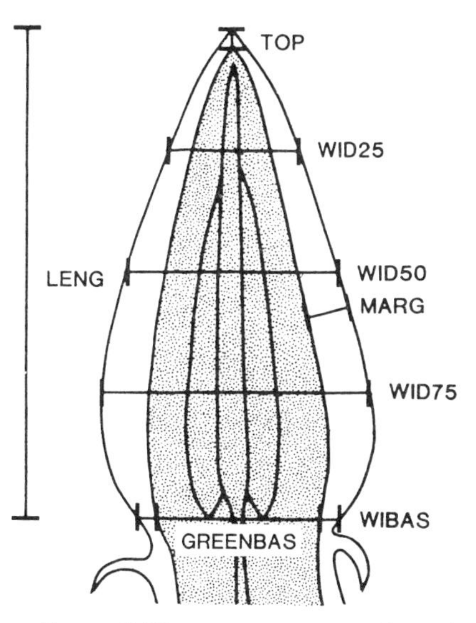

FIGURE 3. The measurements made on the sepal of the flowers of *Scleranthus annuus*.

Variation in Flower Mophology in Natural Populations of *S. annuus*

Forty plants from each of 20 populations in southernmost Sweden (Skåne) were collected at the peak of flowering and transferred to a greenhouse at the University of Lund. In order to study the morphology of the flowers initiated in the field, measurements were made, or flowers were immediately fixed in 70% alcohol.

Two types of data bases were constructed, one with stamen data for the single flowers as entries (3300 cases) and one with plant averages (generally 10 flowers per plant, 331 cases). Sample sizes for the sepal/gynoecium data bases are 3144 and 315, respectively. Variables were log-transformed prior to analysis.

Artificial Crosses

Artificial crosses were made using plants from two types of populations; one population in which the species occurs uninteruptedly over a continuous area, and one in which the species is present in patches (see Svensson, 1988, 1990 for details). The two populations were chosen to represent two extremes of population structure. Populations of *S. annuus* are, due to a high degree of inbreeding and no obvious structures for long-distance dispersal of seeds, assumed to be genetically substructured. Unfortunately, due to the reasons outlined above, no estimates of the genetic structure of *S. annuus* populations is to hand. In the continuous population, relatedness between individuals should decrease with increasing spatial separation. For the population where *S. annuus* occurs in patches, relatedness should be higher within patches than between patches.

Artificial crosses were made in the continuous population using a set of seed parents all originating from a single part of the population. All seed parents used were collected from within a circle with a radius of 0.5 m. These seed parents were crossed with pollen-donors growing at distances of 6 m, 12 m, 25 m, 50 m and 100 m from the seed parent. In addition, selfed progeny was produced by emasculating a plant and pollinating its flowers with its own pollen.

A similar set-up was utilized in a discontinuous population using an isolated patch of seed parents which were crossed with pollen-donors from within the same patch; from several other patches within the population at 7-74 m from the seed parent patch and from another population 300 km distant. Also in this experiment, artificial self-pollination was performed after emasculation.

In both crossing programmes the same seed parent was crossed with several or all different pollen donors, the progeny were sown in a standard environmnet in a greenhouse at the University of Lund and the stamens in 10 flowers from each resulting plant were scored as above and measurements were made on flowers in order to study any effect of crossing distance on flower morphology. Totally 1240 flowers were scored for progeny of crosses from the continuous population, while 2797 flowers were investigated in the crossing programme for the discontinuous population. In addition, data bases with plant averages were constructed, with 123 and 287 cases, respectively. All analyses have been performed on log-transformed data using SAS (SAS 1985) on a Vax computer.

RESULTS

Estimates of Hierarchical Variation in Stamen Fertility Scores and Flower Morphology in Natural Populations of *S. annuus*

Stamen Fertility Scores

The 20 natural populations of *S. annuus* varied in their total flower fertility scores (STAMTOT) from 23.4 to 35.5, with an overall average of 29.5 and comparatively low within-population variation (ANOVA $F_{19,311} = 16.3$, $p < 0.0001$) (*cf.* Fig. 4). Stamen positions in the outer circle of stamens had a general fertility score of 3.2 while the inner circle had lower scores, averaging 0.56 compared with the

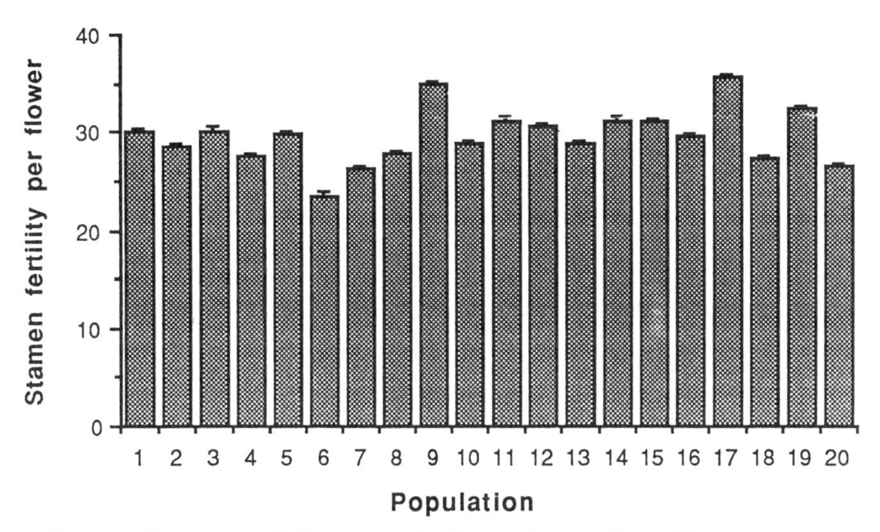

FIGURE 4. Total stamen fertility scores for 20 natural populations of *Scleranthus annuus*. Error bars denote the standard error of means.

outer whorl ($p < 0.001$, t-test) or always carried fully fertile stamens (Table 1). The average fertility score for each stamen position varied among populations, all except stamen positions STA 2 and STA 8 were found to be highly significant in a oneway ANOVA (Table 1). The largest difference for a single stamen position was found for STA 3, which had a score of 1.75 in population #6 and 4.52 in population #17 (data not shown).

The design of the sampling permitted an estimation of the hierarchical distribution of variation to be made. Table 2 shows that 29% of the variation in STAMTOT was among populations, 28% among plants and 43% was within-plants. As can be further deduced, great differences exist between the hierarchical distribution of variation for stamen positions in the outer whorl compared with the inner whorl (excluding STA 2 and STA8), with the outer whorl having a higher proportion of the variation among populations (average 20.4%) than the inner whorl (4.4%).

Sepal and Gynoecium

There is considerable and statistically significant variation among the 20 populations for the 12 characters in the sepal/gynoecium data set (Table 3). The character TOP, which is one of the sepal characters, shows the highest F-value, followed by the two characters representing measurements of the gynoecium. The nested ANOVA, used

TABLE 1. Mean stamen fertility scores for the 10 stamen positions of the flowers of *Scleranthus annuus* (Caryophyllaceae) based on 20 plant averages from each of 20 natural populations. F-values indicate significant differences among populations for individual stamen positions (except STA 2 STA 8) derived from oneway ANOVA (DF = 19, 310). Analysis based on log-transformed values.

Stamen-position	*mean*	*F*	*P*
Positions in the outer circle			
STA 1	3.17	15.87	0.0001
STA 3	3.29	16.74	0.0001
STA 5	3.07	12.04	0.0001
STA 7	3.12	18.00	0.0001
STA 9	3.21	17.17	0.0001
Positions in the inner circle			
STA 2	5.99	0.78	0.7250
STA 4	1.05	5.27	0.0001
STA 6	0.55	5.54	0.0001
STA 8	6.00	0.98	0.4828
STA 10	0.08	1.82	0.0204

TABLE 2. Hierarchical distribution of variation for different stamen positions of the flowers of *Scleranthus annuus* (Caryophyllaceae) at three levels of organization. Values are given as percentage of variation. See text and Fig. 2 for the definition of the variables. Asterisks denote levels of significance for the F-test of MS in the ANOVA, *= $p < 0.05$; ** = $p < 0.001$. POP equals the percentage of variation distributed among populations, IND among individuals, and ERROR the proportion of variation within the individual plants.

	DF	*STAM TOT*	*STA 1*	*STA 2*	*STA 3*	*STA 4*	*STA 5*	*STA 6*	*STA 7*	*STA 8*	*STA 9*	*STA 10*
POP	19	28.8**	19.8**	0	21.5**	8.8**	18.5**	4.9**	20.5**	0	21.7**	0.7*
PLANT	310	28.4**	20.1**	47.3**	20.4**	17.6**	21.5**	12.5**	17.5**	0	19.8**	6.6**
ERROR	2946	42.8	60.1	52.7	58.1	73.5	60	82.6	62	100	58.5	92.7
SUM	3275	100	100	100	100	100	100	100	100	100	100	100

to analyse the hierarchical distribution of variances, indicate that 17–32% of the variation is among populations and 14–50% between plants. In total 31–82% of the variation is among populations and plants, leaving 69–18% as within-plant variation (Table 4).

TABLE 3. Oneway analyses of variance for 12 sepal and gynoecium characters in the flowers of *Scleranthus annuus* plants, testing for differences among populations. 310 plant averages from 20 natural populations were used (DF = 19,291). See Fig. 3 for the definition of variables.

Character	*F*	*P*
LENG	10.10	0.0001
WID25	11.62	0.0001
WID50	11.32	0.0001
WID75	9.36	0.0001
WIBAS	7.97	0.0001
GREENBAS	10.70	0.0001
MARG	10.13	0.0001
TOP	22.91	0.0001
CIRCUM	9.58	0.0001
AREA	10.05	0.0001
STYLE	16.21	0.0001
STIGMA	16.82	0.0001

The Effect of Artificial Crosses on the Variation in Floral Characters

Artificial Crosses in a Continuous Population

The averaged total stamen fertility (STAMTOT) was different for progeny produced by different crossing distances using plant material from the continuous population (Table 5). Crosses made over short distances produced progeny with more stamens and more staminoids with higher stamen fertility scores than progeny produced by selfing and crosses produced over wider distances. The same pattern of response was observed for the summed stamen fertility for organs found in the inner circle of stamen positions in the flower (STAIN) as well as for the variable STAOUT (Table 5). The response to crossing was not equal among the stamen positions in the flowers. Figure 4 indicates an increase in all stamen positions for pooled, crossed progeny compared with selfed progeny, (see Svensson, 1988 for details). The change is, however, not equal among the stamen positions, different stamen positions responding differently to crossing.

Artificial Crosses From a Discontinuous Population

Crosses using plant material from the discontinuous population allowed me to study the effect of four cross-types representing

TABLE 4. Hierarchical distribution of variation for 12 gynoecium and sepal characters among three levels of organization. Values are given as the percentage of variation. See text and Fig. 3 for the definition of variables. Asterisks denote levels of significance for the F-test of MS in the ANOVA, ** = $p < 0.001$.

	DF	*LENG*	*WID25*	*WID50*	*WID75*	*WIBAS*	*GREENBAS*	*MARG*	*TOP*	*CIRCUM*	*AREA*	*STYLE*	*STIGMA*
POP	19	31.5**	28.3**	28.4**	24.6**	20.3**	25.7**	23.1**	16.7**	29.4**	29.4**	28.4**	27.2**
PLANT	291	50.3**	41.5**	42.3**	43.9**	45.06**	39.4**	38.2**	14.1**	50.4**	47.0**	27.8**	22.3**
ERROR	2797	18.1	30.2	29.3	31.5	34.7	34.9	38.7	69.2	20.1	23.6	43.8	50.5
SUM	3107	100	100	100	100	100	100	100	100	100	100	100	100

increasing spatial separation between mates (selfed; within-patch; between-patch and between population) on total stamen fertility. The effect of cross-type on total stamen fertility scores was found to be highly significant (Table 6). Not only did the stamen fertility scores differ among progeny types, but the frequency of fully fertile stamens in the 10 stamen positions was highest in the flowers of progeny produced by the crosses over the shortest distances and decreased with increasing spatial separation between parents (Table 7). Progeny produced by within-patch crosses had 451% more fertile stamens in STA 9 compared with selfed progeny. A relative increase between 325% and 450% was noted for the other stamen positions of the outer circle of stamen postions. Stamen positions in the inner circle showed less increase in the frequency of fertile stamens in crossed progeny compared with selfed progeny. For stamen positions in the inner circle the increase was noted only for within-patch crosses and not for the two other types of crosses (Table 7) Also for this set of crosses, the unequal response among stamen positions is obvious.

In summary, the results from both crossing experiments show that progeny produced by crossing plants separated by short crossing distances have flowers with higher stamen fertility scores compared with progeny produced by artificial selfing. The increase in stamen fertility scores was not equal among the 10 stamen positions.

TABLE 5. The grand mean and sample size of the total stamen fertility scores of all stamen positions of *Scleranthus annuus* (STAMTOT), the summed stamen fertility scores for stamen positions in the inner circle of stamen positions (STAIN) and for positions in the outer circle (STAOUT) presented for each type of cross. Analyses based on plant averages. Data subsets with different letters are significantly different (oneway ANOVA, $p < 0.05$; a posteriori: Duncan New multiple range effects). *Data subset 'crossed'is the pooled values for all separate crossing distances found to be significantly separated from selfed progeny ($p < 0.05$) for STAMTOT, STAIN, STAOUT in separate oneway analyses of variance.

Data Subset	*N*	*STAMTOT*	*STAIN*	*STAOUT*
Selfed	70	25.71^{A}	12.60^{A}	12.35^{A}
Distance 6 m	19	31.81^{C}	14.05^{B}	17.30^{B}
Distance 12 m	5	30.12^{BC}	13.21^{AB}	16.74^{AB}
Distance 25 m	11	27.12^{AB}	12.37^{A}	13.55^{AB}
Distance 50 m	8	27.39^{ABC}	13.00AB	13.79^{AB}
Distance 100 m	11	27.89^{ABC}	13.08^{AB}	14.09^{AB}
Crossed*	54	29.25	13.26	15.22

TABLE 6. Results of analysis of variance for the dependent variable STAMTOT. The analysis was based on plant averages. In this analysis, seed parent patch was defined as a random effect and type as a fixed effect. The interaction between the two effects was used as the error term testing the first effect. Model: $R^2 = 0.114$; $F = 2.32$ ($p < 0.0040$).

	DF	*MS*	*F*	*P*
Type	3	0.07897	6.48	0.0125
Patch	3	0.00230	0.19	0.9012
*Type**Patch*	9	0.01218	1.02	0.4242
Error	271	0.01194		

TABLE 7. The frequency of stamen fertility scores 5 and 6 in the 10 stamen positions in the flowers of *Scleranthus annuus* for different types of crossed progeny, related to the frequency found in selfed progeny. Figures are the percentage increase (*e.g.*, freq. of score *n* in within-patch crosses- freq. of score *n* in selfed progeny/freq. of score *n* in selfed progeny).

	Type										
	STA 1	*STA 2*	*STA 3*	*STA 4*	*STA 5*	*STA 6*	*STA 7*	*STA 8*	*STA 9*	*STA 10*	*Average*
Within-patch (*N* = 325)	450	1	325	289	233	116	447	0	451	17	233
Between-patch (*N* = 829)	354	0	119	-37	108	10	489	-1	451	-8	148
Between-population (*N* = 487)	38	-2	94	89	-17	-50	68	-3	105	-	36

DISCUSSION

Regulation of Stamen Numbers In *Scleranthus*

The study of plant material from natural populations and progeny from artificial crosses using two different populations has shown that *S. annuus* has a variable number of fertile stamens per flower and that stamens instead of staminoids may be found in certain types of progeny. The results presented above also indicate that there is an effect dependent on the crossing distance, *i.e.*, the spatial distance separating the parents of a cross influences the androecium of the progeny.

Price & Waser (1979; Waser & Price, 1983, 1989) claim that if the gene flow in a natural population of plants is restricted and the population is genetically substructured, matings between neighbouring plants should be influenced by inbreeding depression due to the high probability of ancestry by descent. Likewise, matings between parents separated by greater distances should be influenced by 'outbreeding depression' (a fitness decline upon outbreeding) as a result of a disruption of co-adapted genes or of local adaptations (Shields, 1982; Waser & Price, 1989). The combined effects of these two phenomena should favour the existence of an 'optimal outcrossing distance,' progeny from crosses of parents separated by intermediate distances having a superior fitness compared with crosses over shorter and/or greater distances (Waser & Price, 1989). The hypothesis of an optimal outcrossing distance has been tested by several investigators (*e.g.*, Sobrevila, 1988; Svensson, 1988, 1990). Some of the studies support the hypothesis of an optimal outcrossing distance (see review in Waser & Price, 1989) while some could not find any support (Fenster & Sork, 1988; Newport, 1989). Ritland & Ganders (1987) did, however, in a theoretical paper, find support for an optimal outcrossing distance under certain circumstances. A similar conclusion was reached, in another theoretical work, by Uyenoyama (1988).

The results of the two experimental studies on the influence of the crossing distance on stamen numbers reported in this paper indicate that the response of an increase in "stamen fertility scores" declines with increasing spatial distances between the parental plants (*e.g.*, a linear decline reported by Levin, 1984). The shortest crossing-distance in the continuous population resulted in progeny having the highest fertility scores (Table 5), and the comparable least separation in the discontinuous population had the highest stamen fertility scores as well as the highest frequency of stamens classified as fully fertile (Table 7). Thus, in the two studies using populations with different spatial structures, the shortest spatial separation resulted in progeny showing the strongest response to crossing in terms of stamen fertility scores.

This is, to my knowledge, the first study discussing the theory of optimal outcrossing distance for a selfer. Following the assumptions of the optimal outcrossing theory (Waser & Price, 1989) less gene flow would decrease the distance at which the optimum is found (see also Waddington, 1983). Even though a population of a selfing plant species has considerably less gene flow by pollen compared with an outcrossing species, selfers are not devoid of gene flow. The transport of seeds and fruits contribute to the flow of genetic material within populations, although little attention has been given to this topic (*e.g.*, Levin, 1981; Ellstrand, 1992). For the reported experiments showing responses in stamen numbers dependent on crossing distances, the connection with the optimal outcrossing theory is ambiguous. The result can be claimed as support for the theory only if progeny produced by selfing can be regarded as representing the shortest crossing distance.

Non-random Variation in the Androecium of *Scleranthus annuus*

This paper is a report on the results of several studies of the variation of stamen number, both the pattern of variation within and among natural populations of *S. annuus*, as well as the stamen configuration in progeny from artificial crosses of plants from natural, continuous and discontinuous populations. There is remarkable consistency among the different parts of the study. The total stamen fertility of the flowers of plants in natural populations is 23.4-35.5 in the 20 populations studied (Fig. 4). The probability is high that these plants are being produced by selfing (see methods) The two populations used for artificial crosses both have a total stamen fertility per flower for progeny produced by selfing of 25 (25.7 for the continuous population and 25.8 for the discontinuous population). It is clear from the analyses that differences among the 10 stamen positions of the flowers of *S. annuus* occur in a non-random way, *i.e.*, certain stamen positions were more likely to vary than others. Stamen positions in the flowers of *S. annuus* are arranged in two whorls, one inner and one outer (see *e.g.*, Fig. 1). The inner whorl consists of stamen positions that are either fixed in a fully fertile mode (stamen position 2 and 8) or more or less completely reduced (position 4, 6 and 10). The outer stamen positions in *S. annuus* carry staminoids that in this study have been assigned to different stamen fertility scores.

The non-randomness among stamen positions is similar in the studies. The stamen positions in the outer whorl have a stamen fertility score of 3.2 in natural populations with positions 5 scoring lowest, of 2.6 in the progeny of crosses in the continuous populations, and of *ca.* 2.8 in the flowers of the progeny produced by crosses in the discontinuous population. The degree of reduction among stamen positions in the inner whorl that do not carry a fertile stamen is similar in the studies reported. In both studies involving crosses, STA 4 scores lower than STA 6. The most reduced stamen position (STA 10) has very similar general stamen fertility in progeny produced by selfing in the two studies. Stamen positions 5 and 10, being located on the median plane in the inner and outer whorls, respectively, are the positions producing the lowest number of fertile stamens and also staminoids scoring low. Thus, the variation pattern is the same in plants from natural populations as for plants produced by artificial crosses (Fig. 5).

Non-random variation among stamens in flowers has been reported in several instances. Differences between the two whorls of stamens in *Arenaria* (another Caryophyllaceaen genus) were reported by Wyatt (1984, 1988). He observed differencies in anthesis for the two sets in the flowers of *Arenaria uniflora*. The non-randomness in the variation of flower parts has been established previously by Matzke (1932) for plants in Caryophyllaceae. For more general reviews see Bachmann (1983), Hilu (1983) and Durand & Durand (1984). Sterk (1970) found that the sequence of reduction of stamens in *Spergularia* was non-random, certain stamen positions being more likely to be reduced and to lack stamens than other positions. It is clear that the development of the flower, evolutionally a complex structure, does not proceed randomly but follows strictly defined paths.

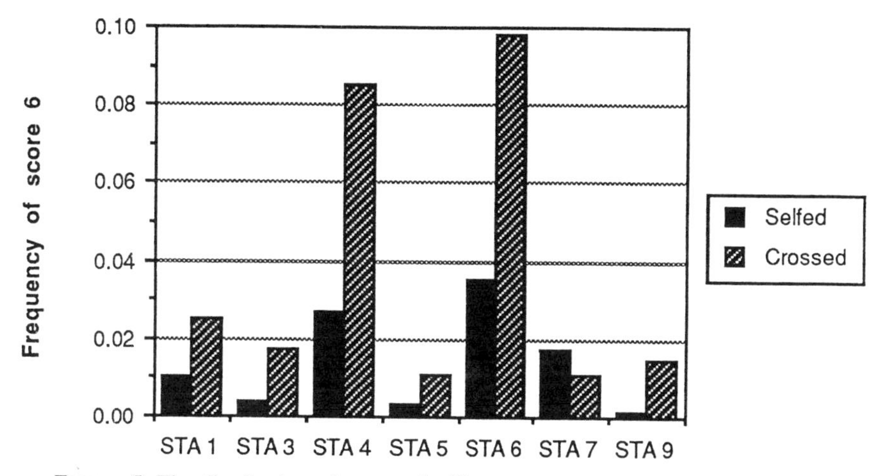

FIGURE 5. The distribution of stamen fertility score '6'among stamens of *Scleranthus annus* for data subset 'selfed'and 'crossed.' STA 2, STA 8 and STA 10 are excluded. See text for further details.

Why is it Possible to Observe Variation in Stamen Numbers in *Scleranthus annuus*?

Why does *S. annuus* exhibit variation in a feature that is closely connected to the breeding system, variation that in outcrossers may influence the reproductive success? (Stebbins, 1950; Seburn, *et al.*, 1990).

The result presented indicates that more than half of the variation found in 'stamen fertility scores,' is between plants and between populations, while about 43 percent of the variation is within plants. Intra-plant variation has traditionally been regarded as environmentally-induced variation, variation frequently treated as error in, for instance, nested analysis of variances (Seburn, *et al.*, 1990; Svensson, 1992). The reason behind the treatment of intra-plant variation may be found in the difficulties in determining whether such variation is adaptive or not (Huether, 1969; Ellstrand, 1983; Seburn, *et al.*, 1990). Half of the variation in stamen fertility scores in *S. annuus* may in part be of genetic origin. This supposition is further strengthened by the observation of experimentally-induced variation in stamen fertility scores in artificial crosses. Svensson & Persson (1994) showed that stamen fertility scores are highly heritable characters in *Scleranthus annuus*. Using 172 maternal families, heritabilities between 0.235 and 0.714 were found for the ten stamen positions. Similar heritabilities have later been found for *Spergularia salina*, another

TABLE 8. Ploidy levels and numbers of stamens reported for some north European members of Caryophyllaceae.

Taxon	*2n*	*No. stamens*	*Source*
Scleranthus perennis	2x	10	Svensson, 1990
S. annuus	4x	2-5	Svensson, 1990
Spergularia media	2x	10	Sterk, 1970
S. salina	4x	5-8	Sterk, 1970
Stellaria neglecta	2x	10	Matzke, 1932
S. media coll.	4x	3-7	Weimarck, 1963

tetraploid annual species of Caryophyllaceae exhibiting varying stamen numbers (Svensson & Persson, in press).

Scleranthus annuus is a tetraploid species with a highly reduced androecium. A typical individual of *S. annuus* in a natural population has two fertile stamens and eight staminoids. *Scleranthus annuus* does not produce any nectar and is normally not visited by any flower visitors. In northern Europe *S. annuus* co-exists with the diploid *S. perennis*, the latter exhibiting an unreduced androecium, *i.e.*, 10 fully fertile stamens, a significant production of nectar and is intensively visited by ants and ladybirds (Svensson, 1985).

Caryophyllaceae is in north Europe represented by several genera with diploid species having 10 stamens and tetraploid species with a reduced number of stamens (Table 8). It can also be seen (Table 8) that a species with a reduced androecium exhibits within-species variation in the androecium. From this, the following hypothetical background is proposed. *Scleranthus annuus* may have its origin in tetraploidization of a diploid ancestor. In connection with the process of becoming tetraploid, the androecium of *S. annuus* may have been reduced and nectar production may have been lost. The reduction of parts of the androecium of *S. annuus* may have proceeded until a steady state of two stamens was reached, the two stamens fulfilling the requirements in terms of pollen grains of a highly selfing annual with a production of one ovule per flower.

The genetic mechanisms for arresting the full development of stamens in all stamen positions in *S. annuus* are, however, unknown and several genetic mechanisms may be involved (see Bachmann, 1983 and Hilu, 1983). The observed variation in stamen numbers in the flowers of *S. annuus* in natural populations as well as in artificial crosses may have its background in an incompleteness in the genetic system arresting the development of stamens. The large variation in stamen fertility in the flowers of progeny produced by crossing plants separated by different spatial distances may also be discussed in terms of incompleteness of the genetic system regulating stamen development. If natural populations of *S. annuus* are genetically spatially structured the variation in the progeny's androecium can be an effect of bringing together genomes that differ to varying degrees.

An alternative way of explaining the observed pattern of variation in stamen number in *S. annuus* is a relaxation of selection on stamen numbers. In a situation with low selection pressure, mutations affecting the fertility of stamens could be maintained in the population, thereby providing a basis for the development of plants with different stamen numbers.

ACKOWLEDGEMENTS

I am grateful to Gunilla Andersson and Helena Persson for technical assistance, Karin Ryde for linguistic advice and financial support from The Swedish Natural Protection Agency and the Natural Research Council. Comments and viewpoints by Dr. Brian Husband were much appreciated and significantly improved the manuscript.

References

Andersson, S. 1989. Phenotypic plasticity in *Crepis tectorum* (Asteraceae). *Plant Syst. Evol.* 168:19–38.

Bachmann, K. 1983. Evolutionary genetics and the genetic control of morphogenesis in flowering plants. *Evol. Biol.* 16:157–208.

Barrett, S. C. H. 1989. Mating system evolution and speciation in heterostylous plants. Pages 257-283 *in* D. Otte & J. A. Endler, eds, *Speciation and Its Consequences*. Sinauer. Ass. Sunderland.

Clausen, J., D. D. Keck, and W. M. Hiesey. 1940. Experimental studies on the nature of species. I. The effect of varied environments on western North American plants. *Carnegie Inst. Washington, Publ.* 520. 450 pp.

Durand, R. and B. Durand. 1984. Sexual differentiation in higher plants. *Physiol. Plant.* 60:267–274.

Ellstrand, N. C. 1983. Floral formula inconstancy within and among plants and populations of *Ipomopsis aggregata* (Polemoniaceae). *Bot. Gaz.* 144:119–123.

Ellstrand, N. C. 1992. Gene flow by pollen: Implications for plant conservation genetics. *Oikos* 63:77–86.

Fenster, C. B. and V. L. Sork. 1988. Effects of crossing distance and male parent on *in vivo* pollen tube growth in *Chamaecrista fasciculata*. *American Jour. Bot.* 75:1898–1903.

Hilu, K. W. 1983. The role of single gene-mutations in the evolution of flowering plants. *Evol. Biol.* 16:97–128.

Huether, C. A. 1969. Constancy of the pentamerous corolla phenotype in natural populations of *Linanthus*. *Evolution* 23: 572–588.

Levin, D. A. 1981. Dispersal versus gene flow in plants. *Ann. Missouri Bot. Gard.* 68:233–253.

Levin, D. A. 1984. Inbreeding depression and proximity-dependent crossing success in *Phlox drumondii*. *Evolution* 38:116–127.

Matzke, E. B. 1932. Flower variations and symmetry patterns in *Stellaria media*, and their underlying significance. *American Jour. Bot.* 19:477–507.

Meagher, T. R. 1988. Sex determination in plants. Pages 125-138 *in* J. Lovett Doust & L. Lovett Doust, eds, *Plant Reproductive Ecology, Patterns and Strategies*. Oxford Univ. Press.

Moore, D. M. 1982. *Flora Europaea: Check-list and Chromosome Index*. Cambridge Univ. Press, Cambridge.

Newport, M. E. A. 1989. A test for proximity-dependent outcrossing in the alpine skypilot, *Polemonium viscosum*. *Evolution* 43:1110–1113.

Ornduff, R. 1969. Reproductive biology in relation to systematics. *Taxon* 18:121–133.

Price, M. V. and N. M. Waser. 1979. Pollen dispersal and optimal outcrossing in *Delphinium nelsonii*. *Nature* 277:294–297.

Ritland. K. and F. R. Ganders. 1987. Crossability of *Mimulus guttatus* in relation to components of gene fixation. *Evolution* 41:772–786.

SAS 1985. *SAS User's Guide, Statistics*, 5 ed. SAS Inst., Inc., Cary, N.C.

Sattler, R. 1973. *Organogenesis of Flowers*. Univ. Toronto Press, Toronto, Canada.

Seburn, C. N. L., T. A. Dickinson, and S. C. H. Barrett. 1990. Floral variation in *Eichornia paniculata* (Spreng.) Solms (Pontederiaceae): I. Instability of stamen positions in genotypes from northeastern Brazil. *Jour. Evol. Biol.* 3:103–123.

Shields, W. M. 1982. *Philopatry, Inbreeding and the Evolution of Sex*. State Univ. New York Press, Albany.
Sobrevila, C. 1988. Effects of distance between pollen donor and pollen recipients on fitness components in *Espeletia schultzii*. *American Jour. Bot.* 75:701–724.
Stebbins, G. L. 1950. *Variation and Evolution in Plants*. Columbia Univ. Press, New York.
Sterk, A. A. 1970. Reduction of the androecium in *Spergularia marina* (Caryophyllaceae). *Acta Bot. Neerl.* 19:488–494.
Svensson, L. 1985. An estimate of pollen carryover by ants in a natural population of *Scleranthus perennis* L. (Caryophyllaceae). *Oecologia* (Berl.) 66:373–377.
Svensson, L. 1988. Inbreeding, crossing and variation in stamen number in *Scleranthus annuus* (Caryophyllaceae), a selfing annual. *Evol. Trends in Plants* 2:31–37.
Svensson, L. 1990. Distance-dependent regulation of stamen number in crosses of *Scleranthus annuus* (Caryophyllaceae) from a discontinuous population. *American Jour. Bot.* 77:889–896.
Svensson, L. 1991. The effect of crossing distance and population subdivision on floral morphology in *Scleranthus annuus* (Caryophyllaceae), a selfing annual. *Plant Syst. Evol.* 174:5–16.
Svensson, L. 1992. Estimates of hierarchical variation in flower morphology in natural populations of *Scleranthus annuus*, an inbreeding annual. *Plant Syst. Evol.* 180:157–180.
Svensson, L., and H. Persson. 1994. Quantitative genetics of stamen numbers in the selfing *Scleranthus annuus* (Caryophyllaceae). *American Jour. Bot.* 81:1112-1118.
Svensson, L., and H. Persson. (in press). Quantitative genetics of stamen fertility in *Spergularia salina* (Caryophyllaceae). *Canadian Jour. Bot.*
Tucker, S. C. 1989. Overlapping organ initiation and common primordia in flowers of *Pisum sativum* (Leguminosae: Papilionoideae). *American Jour. Bot.* 76:714–729.
Uyenoyama, M. K. 1988. On the evolution of genetic incompatibility systems: Incompatibility as a mechanism for the regulation of outcrossing distance. Pages 212-232 *in* R. E. Michod & B. R. Levin, eds., *The Evolution of Sex: An Examination of Current Ideas*. Sinauer Ass. Sunderland.

Waddington, K. D. 1983. Pollen flow and optimal outcrossing distance. *American Nat.* 122:147–151.

Waser, N. M. and M. V. Price. 1983. Optimal and actual outcrossing in plants, and the nature of plant pollen interaction. Pages 341-359 *in* C. E. Jones & R. J. Little, eds., *Handbook of Experimental Pollination Biology.* Scientific and Academic Editions, New York.

Wasser, N. M. and M. V. Price. 1989. Optimal outcrossing in *Ipomopsis aggregata*: Seed set and offspring fitness. *Evolution* 43:1097–1109.

Weimarck, H. 1963. *Skånes Flora.* Corona, Malmö (in Swedish).

Wyatt, R. 1984. The evolution of self-pollination in granite outcrop species of *Arenaria* (Caryophyllaceae). I. Morphological correlates. *Evolution* 38:804–816.

Wyatt, R. 1988. Phylogenetic aspects of the evolution of self-pollination. Pages 109-132 *in* L. D. Gottlieb & S. K. Jain, eds., *Plant Evolutionary Biology.* Chapman & Hall, London.

Aspects of the Genecology of Weeds

Herbert G. Baker
Department of Integrative Biology
University of California, Berkeley, CA 94720

PROLOGUE[1]

In the history of genecology it is notable that the pioneers in this discipline worked exclusively with native plants. Thus, weedy species and others that were introduced into a region by human activity, were ignored. This may be an expression of contempt for non-native species (a prejudice which has long affected authors of floras), but there are also reasons related to the biology and historical treatment. The native species are often seen to occur in disjunct populations (with simple sampling of the 'population' possible) whereas the introduced plants, provided that they are vigorous, may have a more continuous distribution involving complex sampling problems compared with the discontinuous natives.

Motivation for an investigation of natives, rather than introduced species, was also present because the pioneer genecologists were trying to find a taxonomy which reflected ecological characters, rather than only morphological and distributional characteristics. They aimed to recognize the products of ecotypical differentiation as subspecies or varieties. Turesson (1922, 1925, etc.) treated each ecological race as an "Oecotypus." Clausen, Keck & Hiesey (1939, 1940, etc.) equated the ecotype with subspecies in biosystematics. The annual report of these authors working at the Carnegie Institution of Washington was also headed Experimental Taxonomy. Also, in Great Britain, cultivation procedures were known as Experimental Taxonomy. The weeds did not fit into this simple classification which was subsequently shown to be hopelessly inadequate; genecology has

[1] This is not a review but, as its title indicates, it is the consideration of aspects of genecology pertaining to two kinds of colonizing species that we call weeds. The format is to provide one or two examples from the literature for each section. Reviews of the aspects of the wider discipline are provided by Baker (1965, 1972a,b, 1974, and 1991) as well as by Barrett (1982, 1983, 1988, and 1992), Barrett & Husband (1989), Brown & Marshall (1981), Crawley (1986), Harper (1960), Heslop-Harrison (1964), Mortimer (1983), and Rejmanek, *et al.* (1991).

moved away from the classificatory effort and has no need of a taxonomic *raison d'ètre*. Rather, the principles of genecology may be studied in plants that have evolved under human influence just as well as in 'native' species.

There was also a feeling that species introduced by humans would have been exposed to local climatic and edaphic forces for a lesser time than the native species and, for this reason, were less clearly distinguishable as local races. This may be true in some cases but it is not a reason for giving less attention to the adaptive variation in weeds. In California, the majority of plants occurring below two thousand meters elevation are derivatives of man-handled invasions and occur in such numbers that the number of generations is not a limiting factor (Baker, 1989); ecotypical differentiation is there for all to see.

This concern to modify taxonomy by putting it on a genecological basis has largely faded from view as we have come to replace the old alpha taxonomy, which was based on morphology and chorology. Instead we have a synthetic taxonomy which takes into account all of the qualities of the species whether they be morphological, chorological, physiological, biochemical, behavioristic, or cytogenetical (including breeding systems and molecular biology).

Weeds — Ruderals and Agrestals

There are almost as many definitions of weeds as investigators who have studied them, but my own definition, first put forward in 1964, still seems to me to cover the situation most suitably "A plant is a weed if in any specified geographic area its populations grow entirely or predominantly in situations markedly disturbed by man (without, of course, being deliberately cultivated plants)" (Baker, 1965:147). So weeds are colonizing species but with a special connection to human activities that influence nature. They are different according to whether the weeds are *ruderals* or *agrestals*.

Ruderals occur along road-sides and in waste places while agrestals are the weeds of cultivated crops and pastures (Baker, 1965). Ruderals are becoming more common as human population density increases but agrestals, which have been the subject of considerable genecological investigations, are now less common because of sustained efforts at their destruction.

Grime Life History Characteristics

J. P. Grime (1977, 1979) has divided plants three ways, into those which are strong *competitors* as opposed to those which are *ruderals* (with an accent on reproduction) and those which are *stress-tolerators*. All plants have some contribution from each of these proclivities, but they can usually be put in a category according to the strength of their characteristics. There is an overlap in the use of the term *ruderal*. In Grime's treatment 'ruderals' are characterized by strong emphasis on vegetative growth and abundant seed reproduction and this is also a feature of the plants of waste places and road-sides. So there is not much confusion from the dual use of the term ruderal.

Agrestals may be strong competitors or stress-tolerators.

GENECOLOGY OF WEEDS

Race Formation in Weeds

It is apparent that not all ecotypical differentiation results in the formation of clear-cut races (see below), particularly when there is obligate outcrossing and no sharp divisions of climates and soils (Baker, 1954). Where the conditions are suitable, and particularly when the breeding system of the plant is frequent self-pollination, ecological races may be demonstrable (Baker 1954, 1965, 1974).

Matricaria

An interesting example of stable race-formation and migration is provided by *Matricaria matricarioides* (Asteraceae), the visually inconspicuous but often astonishingly abundant 'pineapple weed'. According to investigation by Dr. L. M. Moe (1977) this species and a related species (*M. occidentalis*) are native in northwestern North America. Both are self-pollinating 'treading weeds' (Baker, 1974) and *M. matricarioides* occurs naturally on periodically bare soil such as sandbars whereas *M. occidentalis is* found in vernal pools and swampy grasslands. *M. occidentalis* is not aggressively weedy, but *M. matricarioides* has spread all over North America, Europe, and northern Asia in historical time (*cf.* Hegi, 1929; Salisbury, 1961).

Dr. Moe has been able to demonstrate that there are two separate races in *M. matricarioides*, differing in not very obvious branching patterns, but very obviously differentiated physiologically by the abundant exudation of mucilage when the achenes of one race are wetted and the lack of mucilage production by the other. These races are distinct geographically and the northern race (which does not produce mucilage) is found from British Columbia to Alaska in the New World, and in northern Scandinavia in Europe and Siberia in Asia. The mucilage-producers conform to a more southern distribution in all three continents. Here there is a plausible case of migration on an intercontinental scale that has preserved the features of the races which are completely interfertile and obviously differ in more than just mucilage production.

Capsella

One is reminded of the investigation by G. H. Shull of what he called *Bursa* species (actually races of the autogamous *Capsella bursa-pastoris*, Brassicaceae). Northern and southern races from Europe maintained their morphological and geographical separation, repeating the European geographical pattern on introduction (probably many times more than once) to western North America (Shull, 1937).

A modern investigation of *Capsella bursa-pastoris* utilizing seeds collected over a 5,000 km transect from Scandinavia to the central European Alps was presented by Bosbach & Hurka (1981). They do not report morphological or physiological characteristics of the populations but do report allozymic analyses, finding a correlation between the amount of disturbance in the habitat and the heterozygosity of the allozymes. These were more numerous in heavily disturbed habitats whereas lightly disturbed habitats might only have one allozyme. The significance of this finding remains to be seen.

Population distinctness on the microscale is shown by grasses; thus the autogamous annual ruderal *Poa annua* in a mosaic environment of bowling green lawns and flower beds in Cambridge. (Warwick & Briggs, 1978).

The *Matricaria* and *Capsella* examples are ruderals, and the influence of their habitual self-pollination on the geographical picture should be considered. They can produce a self-reproducing colony from a single seed in a suitable habitat enabling long-distance seed dispersal (Baker 1955, 1974). But race-formation is also to be seen in autogamous taxa among the agrestals.

Flax (*Linum*) and *Camelina*

The classic case of close ecotypical adaptation of a weed to the crop plant with which it grows is provided by *Camelina sativa* agg. (Brassicaceae) in flax (*Linum usitatissimum*, Linaceae) crops of central and eastern Europe and southwest Asia. The best analysis of the work carried out on this association by Russian, Polish, and Swedish scientists in the first three decades of the twentieth century is by G. L. Stebbins in his book *Variation and Evolution in Plants* (Stebbins, 1950).

Where cultivation of flax is irregular or dependent on primitive scratching of the ground, weedy *Camelina sativa* is variable in morphology, physiology, and phenology. It is a winter annual, but easily transformed by selection into a summer annual. However, in areas where flax is grown predominantly for stem fiber, the flax plants have been selected for a tall, almost unbranched stem that casts considerable shade in the summer. They represent a very competitive crop for any weed to deal with. However, in these crop fields, the weedy *Camelina sativa* is equally unbranched and bears its flowers at a height corresponding to that of the flax flowers. It breeds true because it is self-pollinating. There is some evidence that the direct effect of the shady environment may influence the *Camelina* plants to some extent, but cultivation experiments show that the form of the plant is genetically determined, and presumably resulted from 'natural' selection. Not only is the morphology of the *Camelina* plant adapted to life in the crop field but its flowering is timed to agree with that of the crop plant and as a result, when the flax seed is harvested, the seed of *Camelina* is taken too. When the seed sample is purified of fruit pod parts with their removal by winnowing, the *Camelina* seed reacts to the gusts of air similarly to the seed of *Linum*. Consequently the *Camelina* seed gets sown with the crop plant seed next year. It fully deserves the name var. *linicola*.

Other forms of *Camelina sativa* are adapted to other crops. Thus, a widely branched oil-seed form of flax (*Linum crepitans*) with easily shattering pods is grown in some areas of Russia. It is accompanied by an equally branched, equally easily shattering form of *Camelina* (called *C. crepitans*). Both are harvested just before the pods open.

Other species, such as *Silene linicola* (Caryophyllaceae) infest flax fields but not so constantly as the flax-adapted ecological races of *Camelina sativa*.

Rice (*Oryza*) and *Echinochloa*

Rice (*Oryza sativa*) has been grown commercially in California since 1912. During this time 62 weed species have been recognized in an aquatic weed flora of the paddies (Barrett & Seaman, 1980). These were largely recruited from irrigation canals, as well as pre-existing rivers and ponds. An example of race formation in this environment by a crop weed is illustrated by a pair of species of *Echinochloa* (Poaceae), *E. crus-galli* agg. and *E. oryzoides*. *E. crus-galli*, often referred to as barnyard grass, has been a common weed in swampy ground in California for a long time. But it appeared as a serious rice paddy weed in the California fields. Because of its autogamy and six sets of chromosomes, the taxonomy of this species is very difficult. However, a clear distinction between it and another polyploid, *E. oryzoides* (frequently referred to in literature as *E. crus-galli* var. *oryzicola*), can be drawn. *E. oryzoides* almost certainly came to California fields in contaminated seed stock from eastern Asia. We owe the story of this interaction in California to Spencer Barrett (1992).

The most common form of barnyard grass is *Echinochloa crus-galli* var. *crus-galli* which is found on paddy bunds and in shallower parts of the paddy itself. *E. oryzoides* is morphologically, physiologically, and phenologically distinct, and its tetraploid form is found only in the paddy itself. It is only known from rice fields in different parts of the world.

Traditionally, rice fields are flooded and seedlings are raised in a nursery and planted out in shallow water. The water is increased in depth as the season goes on. In California, the germinated seed is now sown from airplanes in April, and there used to be a draw-down of the water for a couple of weeks after sowing the germinated seeds, allowing the rice seedlings to root themselves. But this also lets seeds (from the soil) of *E. crus-galli* germinate (which they cannot do under deep water) and become established as a weed in the paddy. California farmers did away with the draw-down and so control the establishment of the population of *E. crus-galli* in the paddy. But this lets the *E. oryzoides* (which *can* germinate under deep water) establish in its place.

These two closely related *Echinochloa* species also differ in their maturation-rates and flowering times. The early-flowering *E. crus-galli* (it begins flowering in June) is morphologically distinct from

Oryza, with floppy leaves and red pigmentation and sheds its fruits before the rice is harvested in September. *E. oryzoides* plants are dark green and have upright leaves; in fact, they are very similar to vegetative rice plants. They flower at such time that their seeds may be harvested with those of the crop, although a seed bank is also developed from fruits that fall during the harvest period. They are an example of a 'crop-mimic.'

There is no evidence that the *E. oryzoides* form has arisen in California in the past few years; it has been recruited from contaminated seed stock from other parts of the world.

Barrett (1983) considers the vegetative mimicry of rice by the plant of *E. oryzoides* to have resulted from selection of the species against hand weeding on a visual basis in the intensely managed rice culture of the Far East. If this be the case, the vegetative crop mimicry has been joined with the timing of fruit maturation to provide a very important weed as a result of crop-handling techniques. For 'high tech' commercial agriculture, a very important weed race was preadapted to modern techniques. The purity of the weed is preserved by self-pollination.

A review of crop mimics in general is given by Barrett (1983). The evidence is strong that they are closely adapted to the environmental circumstances in which they occur and make very strong competition for the crop plant. They are self-pollinating and genetically relatively uniform as indicated by the results of electrophoretic analysis (Barrett, 1992). A recent switch to 'dwarf' rice cultivation has enabled farmers to recognize the mimic, and make it possible to eliminate it.

General Purpose Genotypes

In 1964, I suggested another strategy which may account for the success of weedy plants. Here, instead of race-formation in response to local selection, the individual plant is adjusted to a variety of environments through the possession of a genotype which has this capacity. I call this possession by *individual* plants of broad ecological tolerance the expression of 'general purpose genotypes' (GPGs) (Baker, 1965).

An example of a plant with a GPG strategy is tetraploid *Ageratum conyzoides* (Asteraceae), which is one of the most common weeds in tropical, subtropical, and warm-temperate regions of the world. A remarkably close approximation to freedom from control by abiotic factors of the environment is shown by individuals of this species (Table 1). This freedom is especially seen by contrast to the narrow environmental tolerance of individuals of the scarcely weedy *A. microcarpum*, a probable diploid ancestor of *A. conyzoides* on the Central American plateau. Individual plants of *A. conyzoides* show their freedom in germination, vegetative growth, flowering, and seed setting.

TABLE 1. Comparative features of *Ageratum microcarpum* and *A. conyzoides* revealed by controlled environment experiments.

Ageratum microcarpum (scarcely weedy)	*Ageratum conyzoides* (widespread weed)
Light requirement for germination	No light required for germination
Perennial	Life span year
Flowers in second season of growth	Germination to flowering in 6-8 weeks
Flowering inhibited by high night temperatures	Flowers at low (10°C) or high (27°C) night temperatures
Flowering better with long (12 hour) nights	No photoperiodic control of flowering
Mesophyte	Tolerates waterlogging, drought
Self-incompatible	Self-compatible (largely self-fertilized)
Not very phenotypically plastic ($n = 10$)	Phenotypically plastic ($n = 20$)

A notable feature of most weeds, especially annuals, is their ability to set seed without the need for pollinator visits, either by autogamy (self-fertilization) or agamospermy (formation of seed without the sexual process). Even when outcrossing does take place, wind or generalized flower visitors are utilized. The advantages of autogamy or agamospermy for a weed include provision for starting a seed-reproducing colony from a single immigrant or regeneration of a population after weed-clearing operations have removed all but a single plant. Such breeding systems allow rapid build-up of the population by individuals virtually as well adapted as the founder. This combination of the GPGs with self-compatibility allows for opportunism such as is characteristic of weeds.

Where the weed is a perennial, self-compatibility is less certain to be found (and some perennial weeds are even dioecious), but an extra emphasis upon vegetative reproduction here achieves the same end, that is the rapid multiplication of individuals with appropriate genotypes.

Many weeds have outcrossing relatives that probably reveal the ancestral condition. Genetical recombination in such outcrossing ancestors may provide the appropriate general purpose genotypes that may then be replicated by autogamy, agamospermy, or vegetative reproduction.

It is not intended to imply that there is just one general purpose genotype for each species; undoubtedly there will be many combinations of genes that will provide the wide range of ecological responses and some may be more general purpose than others. It seems that even if general purpose genotypes are emphasized in the colonization, there will be a tendency for the selection of locally adapted genotypes as the species persists in an area. This has been postulated for *Taraxacum officinale* (Baker 1965, 1974). This area is liable to be altered considerably.

In all cases the weeds show phenotypic plasticity in the face of environmental variation, being able to reproduce even if their vegetative growth is diminished by much less than optimal conditions. The preservation of genetical variability, as well as of the population as a whole, must be assisted by such plasticity (Bradshaw, 1965; Jennings & Trewavas, 1986; Stearns, 1989).

A similar theory of plant responses to the imposition of stress factors has been put forward very recently by Chapin, *et al.* (1993). This is a centralized stress response system (RGR). Evidence for it is based on experimental studies in which *similar* physiological changes are observed in response to manipulation of a *variety* of stress factors. "Those changes that had a positive effect on fitness would be selected and remain in the population. Most research on the centralized stress response system has been done on crop plants, so there is currently no evidence as to whether ecologically distinct species differ in hormonal responses to environmental stress" (Chapin, *et. al.*, 1993:S83). The influence of these pauci-genic systems is ideal for incorporation into General Purpose Genotypes (GPGs).

Eupatorium adenophorum

In the genus *Eupatorium* (Asteraceae) the most strikingly successful weed is of this nature. It is *Eupatorium adenophorum* Spreng. (also known in the literature as *Eupatorium glandulosum* HBK.). As Holmgren (1919) first showed (and I have confirmed), plants of this species are triploid (with 51 chromosomes) and form seeds without pollination. Embryo-sac formation follows diplospory and the unreduced egg nucleus then develops by diploid parthenogenesis. The abundant tiny achenes produced by this agamospermous process serve for rather long-distance dispersal, while consolidation of the occupancy of an area is achieved by seed and by vigorous vegetative reproduction.

Such is the range of environmental tolerance in this species that it is well established from above 2,500 meters in its native Mexico to tropical Jamaica and the subtropical rain forests of Hawaii, and in California from the aridity of the canyons near Pasadena to the foggy coast near San Francisco Bay. It is a serious weed in Australia, Hawaii, India and in the Canary Islands. It has been tamed by biological control methods in Hawaii and Australia by using a stem-gall fly (Wilson, 1960). Its remarkable spread has been achieved even though genetical recombination must be at least very much slowed down by agamospermy. Plants raised in the greenhouse in Berkeley from seed collected in Mexico, in Hawaii, and in Marin County, California were extremely similar in relation to moisture and temperature reactions. In all probability, *Eupatorium adenophorum* operates on a GPG strategy.

Cortaderia

In the genus *Cortaderia* (Poaceae) in California there is a particularly apt contrast between the life-histories of two perennial species of ornamental giant grass (*C. jubata* and *C. selloana*) (Costas-Lippmann, 1976). *C. jubata* is a native of the Andes in Bolivia, Ecuador, and Peru whereas *C. selloana* is the 'Cortaderia selloana-pampas grass' from Argentina. *C. jubata* was mistakenly identified as *C. selloana* in Munz's (1963) California Flora. *C. jubata* is also known as a weed in New Zealand (Connor, 1973; Connor & Edgar, 1974). Experimental and observational investigations by Costas-Lippmann (1976) show the strong contrast between these species when they are given the opportunity of escaping to life as weeds.

Cortaderia jubata has become established in habitats such as eroded man-made embankments, cliffs, and cut-over areas in the redwood forest, particularly in the coastal foggy belt from Monterey County to Humboldt County and, less frequently, in open habitats further south in California. Weediness in *Cortaderia jubata* is favored by comparable germination capacity of seeds from contrasting habitats and a 'continuous' germination pattern (with no light requirement). A striking feature of *Cortaderia jubata* is its ability to germinate and establish in a variety of soils including those derived from sandstone, shale, rhyolite, quartz-diorite, and serpentine. This catholicity in success in these varied soils is impressive because of the presumably genetical identity of the seeds as a result of their apomictic production.

A two year vegetative phase in development of *C. jubata* from seed is followed by flowering, with the tendency to double flowering during the same season, development of 100% of the ovaries into viable caryopses through apospory and diploid parthenogenesis. Its potential for spreading in California is limited by the need for bare areas for seedling establishment, for the absence of a frosty winter and the presence of a foggy summer in coastal areas.

By contrast, in *Cortaderia selloana* we see an abrupt pattern of germination (which has a definite daylight requirement), lack of ability to establish seedlings in at least serpentine soils, a longer vegetative phase, a shorter flowering period, an inability to flower for the second time during the same flowering season — all diminishing its potential as a colonizer. Also, it needs a plentiful water supply. Its near failure as a weed appears to be related to the consequences of its dioecious breeding system (Costas-Lippmann, 1976). Few gardens have room for more than one pampas grass plant, so its seed production is virtually impossible.

But even *Cortaderia jubata* is not an ideal weed for it has climatic restrictions and its seedlings cannot stand the competition of lowly grasses in closed grassland or lawns. This appears to be due to their small size and the small amount of nutrients in each seed. But the fantastic output of seeds, all of comparable genetic constitution (the consequence of the apomictic process), label it as a GPG-strategist and its pre-adaptation to life in a variety of soil types is striking.

If weeds can make good use of 'general purpose genotypes,' why are similar GPGs not prevalent also in the nonweedy plants which have been investigated by genecologists?

GPGs in Non-Weeds

Actually, Clausen, Keck, and Hiesey's experiments in California have shown that this kind of genotype can be produced by recombination following the crossing of ecological races in *Potentilla glan-*

dulosa Lindl. (Rosaceae) (Clausen, 1949). When the F_2 from a cross between a foothill race and a subalpine race of *Potentilla glandulosa* was examined by growing ramets at Stanford (30 m. elevation), Mather (1,400 m.) and Timberline (3,050 m.), there were some ramets which did best at one of the stations and some ramets which were weak at all stations; but there were some ramets which grew well (in the absence of competition) at all three stations. In their writings, Clausen, Keck, and Hiesey do not comment on the ecological or evolutionary significance of these successful plants; I should say that they possessed 'general purpose genotypes.'

Similarly, in the genus *Poa* (Poaceae), following the artificial crossing of two facultatively apomictic taxa, *Poa ampla* Merr. and *Poa pratensis* L. var. *alpigena* E. Fries, a new strain of plants, also highly fecund and apomictic, was established experimentally. It was unusually tolerant of wide differences in climate (Clausen, 1953). It grew vigorously in experimental gardens at latitudes ranging from San Fernando in southern California (at 34° N) to Volbu in Norway (situated at 61° N and at 45 meters altitude). As Clausen (1953) wrote, "Such wide ranges of tolerance are not found in any other *Poa* in the wild." It must be emphasized that these experimentally produced forms with 'general purpose genotypes' were never exposed to competition with any other kind of plant in a natural environment.

Generally, however, it seems that for nonweedy species, and particularly for those which grow in closed communities, such as perennial grasslands, adaptation through races of relatively limited tolerance but exquisite local adjustment is usually necessary for success in intense competition. The 'general purpose genotypes' which can be selected artificially may not be selected naturally in *closed* communities, for they produce plants which tend to be 'Jack-of-all-trades-but-master-of-none.'

It is European experience that weed species seem most often to have been recruited from open habitats — seashores, the margins of salt marshes, rocky places in mountains, river banks, and the like. Harry Godwin (1960) recorded the achenes of *Taraxacum* in Arctic tundra of the Pleistocene. These are all habitats where 'general purpose genotypes' and pioneering ability might already have proved advantageous.

It is most likely that GPGs are selected from environments that are unpredictable. This is likely to be the case when dispersal and colonization are of importance in non-weeds as well as weeds.

Agrostis capillaris

A very recent series of papers by Rapson and Wilson documents the results of very careful investigations of the genecology of *Agrostis capillaris* introduced into New Zealand some one hundred and fifty years ago. This species, which has appeared frequently in the literature as *Agrostis tenuis* (a name that is now outmoded), has been worked on by resident and guest population biologists at the University College of North Wales, Bangor, Wales. It is famous for its evidences of rapid race formation in response to heavy metals in the soil (Antonovics & Bradshaw, 1970; Antonovics, *et al.*, 1971) and also to crowding on conventional soils.

But Rapson and Wilson found evidence of distinctly nonadaptive features in the genotypes of the plants growing in New Zealand habitats. Rapson & Wilson (1988) sampled genotypes as tillers from populations over a wide environmental range in southern New Zealand and transplanted them back into their own and each others' sites. Growth, phenology, and tiller population dynamics all gave very little evidence of adaptation at the population level despite the successful invasion of the sites during a century and a half. This introduced dryland species was then examined in greater detail by the same authors. Phenological studies showed no significant differences between population samples in culture. (Rapson & Wilson, 1992a). Another paper by the same authors (Rapson & Wilson, 1992b) on responses to light, soil fertility, and water availability showed some evidence of genecological differences in response to marked site differences in soil and water availability. To quote these authors (Rapson & Wilson, 1992b:13),

> Population-based variation in specific leaf area was correlated with rainfall and soil nitrogen and phosphate levels. Differences between populations in other characters did not correlate with environmental trends. Responses indicate some adaptation of populations in their local environments in New Zealand, but also high within-population variation and some instances of nonadaptation. Plasticity is high — as much as 80% of the total variation between treatments — as is typical for a weedy, invasive species. Plasticity, coupled with some genecological differentiation, seems the most likely explanation for the widespread distribution, both spatially and ecologically, of *Agrostis capillaris* in New Zealand.

This naturalized grass has been in New Zealand for about 150 years and it would be interesting to see if the adoption of local races has achieved greater dimensions in those areas which were first invaded compared with those areas in which invasion is still going on. 'Plasticity,' Rapson & Wilson (1992a,b) suggest is the alternative means of evolution for such an indeterminate perennial weedy species. To quote further from Rapson & Wilson (1992b:21):

> Plastic responses involve perception of an environmental stimulus and the initiation of the appropriate response. Plasticity acts as a block to evolution by hiding genotypes under the temporary manifestation of a suitable phenotype. Nevertheless plasticity is the ideal means of adaptation of long-lived individuals under which category clonal grasses may be included and is of particular importance in environments which fluctuate in space or time, the common habitat of weedy or invasive species.

I suggest that general purpose portions of the genotype confer the plasticity in this picture.

The mixture of general purpose and local adaptation genes may varied in each direction. Partial GPGs may represent stages in the breakdown or buildup of full-scale GPGs, or of races. But it is clear that we should not accept that every gene in a genotype is working for the local race in the genecology of colonizing species. We ought not to feel too badly about evidence of nonadaptive genes in natural or unnatural populations. For even if a genotype gives only 90% adaptation to environmental circumstances it will triumph over one that has merely 80%.

The replacement of general purpose genotypes, which may have been responsible for invasion of a new habitat, by locally adaptive genes, is not likely to take place in a single step. Consequently, we may expect to find further evidences of both GPG strategies and local adaptation genotype strategies in the same taxon.

Typha latifolia

Possibly this heterogeneity is to be seen in the aquatic cattail, *Typha latifolia,* a species which has been shown to have distinct climatic races in North America (McNaughton, 1966, 1970, 1974). Nevertheless, this species colonizes acid mine-drainage in Pennsylvania and otherwise de-vegetated areas near smelters in Sudbury, Ontario. In a collaboration between eight authors (McNaughton, *et al.* 1974), no genotypic differences in growth rate could be seen in their experiments between material originating in a non-heavy metal soil from material growing in heavily contaminated areas. These authors suggest that there is some peculiarity of the physiology of *Typha*, in comparison with other plants studied, with a general resistance to heavy metals (McNaughton, *et al.*, 1974). To me this suggests the operation of both race formation and GPG strategies in *Typha latifolia.*

Importance of the Breeding System to Differentiation

Göte Turesson was conscious of the importance of outcrossing in producing seed in self-incompatible plants, but he did not test his material for self-incompatibility as a general rule. Clausen, Keck, & Hiesey (1939, 1940, etc.) regularly tested their material for its breeding systems. All genecologists have appreciated a greater or lesser effect of breeding systems on ecotypical differentiation, but few of them have placed it in the first position in this respect. This view may change.

Outcrossers and Selfers

The ability of a plant to reproduce after 'long-distance' dispersal is very positively increased if the plant can set seed by selfing or apomixis (Baker, 1955, etc.). A self-incompatible plant must wait until another, hopefully compatible with it, arrives and grows (Baker & Cox, 1984). Even if this lucky event takes place, there is always the danger of inbreeding depression in the new population that is forming. This inbreeding depression is particularly likely in the case of plants that are self-compatible but generally outcrossing in the original habitat. In addition, outcrossing species may require specialized pollinators, which may not be available or attractable to the flowers of a single plant or a few plants.

Dioecism

Although most weeds are probably self-pollinating or apomictic, there are successful examples of weed species that are even dioecious. An example is *Silene alba* (Caryophyllaceae), which is aggressive in Europe. It has become successfully introduced into North America by means which take care of the dioecism. Thus, it was introduced through ships' ballast heaps in the vicinity of Philadelphia (Baker, 1986). The earthen ballast was shoveled in from waste land in Europe and shoveled out again when the holds of the ships were to be stocked with American commercial products bound for Europe. Each ballast heap contained a multitude of seeds which were capable of producing a multi-plant population with representation of both staminate and pistillate plants. It was able to spread further by occupation of pastures and cultivated fields. It is primarily moth-pollinated and attractive enough to succeed in competition with native moth-pollinated species for the attentions of the pollinator.

Another successful weed which is dioecious is the 'Canada thistle' (*Cirsium arvense*) which has spread to California from the north, aided chiefly by its vigorous vegetative reproduction together with some imperfection in the dioecism.

Apomixis

Many weeds are apomictic. They reproduce by means that do not involve sexual processes. The result is the same as with vegetative reproduction: production of plants that are genetically identical with each other and with the parent plant. Apomixis can be a very valuable characteristic for a species to build up a population after arriving as one or a few seeds. There is an advantage over self-pollination in that there is no inbreeding depression. Some of the apomictic weeds are uneven polyploids, which would be sterile were it not for the apomictic process.

Several apomictic mechanisms are to be seen in weeds: apospory plus diploid parthenogenesis (particularly in grasses); diplospory plus diploid parthenogenesis (*cf. Taraxacum* spp.). There may be a necessity for pollination for the formation of endosperm (pseudogamy). A replacement of flowers by bulbils is also to be found (*e.g.*, in some grasses). The different methods vary in the extent to which they are obligate, and facultative apomicts have greater evolutionary potential than obligate apomicts.

Apomixis may be a protector of general purpose genotypes from genetic breakdown by recombination.

Pseudogamous species which need pollination to provide for endosperm production are less able to set seed in conditions unfavorable for pollinators, but they are also to be found among the weeds. An example is *Hypericum perforatum* (Hypericaceae). This very successful weed in western North America, South Africa, and Australia forms dense populations with flowers that are attractive to bees, and the necessity for pollination does not appear to have hindered this species (Pritchard, 1960).

Cloning Reproduction

Cloning, the vigorous vegetative reproduction of a parent plant, is especially notable in the weeds of two strikingly different ecosystems, *viz.* perennial, grazed grasslands, and slow-flowing or still water.

The grasses of the perennial, temperate, grazed grasslands (with summer rain) are generally self-incompatible which makes for selection of fine adaptation. Good examples are provided by species of *Agrostis, Festuca*, etc. Another clonal grass which is self-incompatible when sexual is *Poa pratensis*, which has the great advantage of reproducing vegetatively, agamospermously, and sexually by outcrossing. It is a bold competitor.

The aquatic habitats have seen some of the most spectacular spreads of weeds, including such notorious examples as tropical and subtropical *Eichhornia crassipes* (Pontederiaceae), temperate *Myriophyllum spicatum* (Haloragaceae) and *Elodea canadensis* (Hydrocharitaceae), as well as the tropical aquatic fern *Salvinia molesta.* These species rarely reproduce by sexual means, although when this does occur, genetic recombination of limited extent is possible. *Elodea canadensis* managed to block waterways in the British Isles, even though only female plants were introduced from North America.

Barrett (1980) has observed the circumstances in which seed production and germination can occur in *Eichhornia crassipes* despite the prevailing vegetative reproduction. Geber, *et al.* (1992) have shown the outcome of this, that there are genetic differences in clonal demography in *E. crassipes* from different localities in Mexico.

The suspicion that all these exploding species are doing it with general purpose genotypes is very strong.

An unassailable example is the vegetatively exuberant *Oxalis pes-caprae* (Oxalidaceae) from South Africa, which has become a noxious weed in California, Europe, and Australia (Baker, 1965, 1974). This pentaploid species never sets seed, but reproduces vigorously by formation of offsets subterraneally. *Oxalis pes-caprae* will out-compete any grass but, although it flowers freely, it is seed-sterile.

Grasses of the genus *Agrostis* have played an important role in the grazing economy of the grasslands of the Thames Valley in England. A. D. Bradshaw (1959, 1960) has claimed that because of grazing, these grasses almost never had a chance to flower and set seed. In these circumstances, hybrid vigor is advantageous but paucity of seed production is not disadvantageous. Clonal growth is maintained and the grasslands contain high proportions of hybrids between *Agrostis stolonifera* and *A. capillaris* (*A. tenuis*). Temperate grasses may cover large areas with a single clone.

Tropical grasses often have large weedy potentials and a whole paper might be devoted to their weedy behavior in the formation of 'new grasslands' in the American tropics (*cf.* Baker, 1978). *Pennisetum clandestinum* from East Africa and *Cynodon dactylon* from the Middle East or Africa are perfectly adapted to grazing pressure from their experience with large ungulate populations in East Africa. They reproduce freely by stolons and infrequently by seed (Baker, 1978). This enables them to colonize areas from which woody cover has been removed. They have general purpose genotypes which enable them to colonize weedily far away from the tropics, even northern California, although they are not perfect examples because they are frost-sensitive, so that the interplay of GPG and a locally disadvantageous physiological character is also apparent.

A similar pattern of reproduction is shown by numerous herbs or forbs. *Ranunculus repens* (Ranunculaceae) has become a weed of grasslands in temperate zones throughout the world (Sarukhan, 1974).

Variation in Breeding Systems

Clearly it is essential, in any investigation of ecotypical differentiation, that as much as possible be found out of the reproductive biology, including the breeding system, before conclusions are drawn about the evolutionary processes going on.

It should always be remembered that breeding systems can change. An example is provided by *Picris echioides* (Asteraceae) which appears to be native on sea cliffs and salt marsh-margins in Europe. It is perennial and decumbent, but it has given rise to the form which grows as a weed in cereal fields and waste places. This form is upright, annual, and self-compatible. It has spread as a ruderal in North America (Baker, 1974).

Also sexual and apomictic races of the same species may exist and embryos of sexual and asexual production may be present in the ovules of the same plant (as in facultative apomicts; *e.g. Poa pratensis*).

Measurements of morphological characters in populations show a greater tendency for variation in outbreeding species (Hamrick, *et al.*, 1979). However, it is also clear now that we must not consider even habitually selfing species as showing no genetic variation in their populations. The research of Robert Allard and his associates and Subodh Jain at the University of California, Davis, on the weedy grasses *Avena barbata* and *A. fatua*, fit with this evolutionary scheme. *Avena barbata* seems to operate an approach to a 'general purpose genotype' strategy (and has considerable developmental flexibility to go with this); *Avena barbata* shows lesser genetic differentiation into local and microenvironmentally-adapted races than *A. fatua*, although at least two major genotypes are involved (Allard, *et al.*, 1972, Hamrick & Allard, 1972). Heterozygosity can also be maintained in species that are reproducing vegetatively or by apomixis. Allopolyploids may have stable *intergenomic* heterozygosity.

Armeria

In Europe, *Armeria maritima* (Plumbaginaceae) is an inhabitant of sea cliffs and maritime rocky places. It also occurs inland in rocky mountainous areas in soils with a higher than usual heavy metal content. Since the creation of lead and zinc mines in Belgium, *Armeria maritima* has colonized the spoil-heaps (Lefèbvre, 1970, Lefèbvre & Vernet, 1976, Vekemans, *et al.*, 1990). Although the populations of *A. maritima* on the mine-heaps are distylous (with the populations containing both self-incompatible morphs), there appears to be a softening of the self-incompatibility in one of the forms, with 30-40% seed-setting following self-pollination. Lefèbvre attributes this to selection for colonizing ability on new mine-heaps ('mine-islands') but with the advantages of outcrossing when the population is established.

It is a different story in the Arctic, as well as in populations of *Armeria maritima* all down the Pacific Coast to Southern California and, jumping the tropics, along the Pacific coast of Chile to Patagonia and Tierra del Fuego (Baker, 1953, 1966; Vekemans, *et al.*, 1990). Here the dimorphic incompatibility system has been overcome and the species is naturally self-fertile through derivation of a monomorphic, completely self-compatible, morph by combining the pollen character of one morph with the style of the other morph. This seems to be a response to difficulties in the Arctic of pollination which has since been beneficial to *Armeria maritima* as a colonizing species in the New World (Baker, 1953, 1966).

Seed vs. Ramet

As experimental studies of genecology have matured, they have demonstrated greater and greater subtlety of genetically determined variation and adaptation. Although it may be correct still to regard the products of cross-pollination as being more variable than those that follow from self-pollination and, even though asexual processes produce the seed in apomicts, genetically slightly different seedlings may come from them all. With the prodigious production of seeds by many colonizing species, this may result in the recombination of these differences in sufficient numbers to be evolutionarily significant (but see p. 193).

Consequently, it may not be proper to draw conclusions from plants grown in a 'uniform garden' without distinguishing between garden populations raised from seed collected in the wild (genets), and those raised from ramets collected in the wild. The first gives some indication (if enough plants are raised) of the potential for future populations, while the latter illustrates the structure of the sampled population.

Turesson transplanted almost all plants for cultivation as individuals. By contrast, Clausen, Keck and Hiesey began all their cultivations from seeds.

Ramets as well as seeds may convey influences of their past environments. In perennials, the architecture of a plant that is transplanted to a garden from a state of nature may persist for several years.

Quinn & Colosi (1977) reviewed the situation and made suggestions for obtaining comparability and generating information about the range of potentially adaptive responses in what they refer to as "separating genotype from environment in germination ecology studies."

The same problem was tackled by Hume & Cavers (1981) in an interesting study of *Rumex crispus*, a widespread invader of North America from Europe. They raised plants from both achenes and ramets in a uniform garden in Ontario, Canada, using material derived from two populations occurring in contrasting habitats: river flood-plain and grazed pasture. Fifty-eight characters were measured on each individual plant in the cultures. The flood-plain population is inundated annually in a physically more severe annual cycle than endured by the pasture population. The latter, however, is subjected to grazing pressure.

In the case of the pasture cultures raised from seed and ramets, there was a significant difference in variability in favor of the seedlings. There was a significant difference when the flood-plain and pasture cultures were compared; the flood-plain cultures had greater variability in both seedling cultures and ramet cultures than the pasture cultures, and all cultures showed more variation in the seedling cultures than in the ramet cultures with more significant differences in the seedlings' offspring than in the ramets' offspring.

Direct Influences of the Environment

When Turesson observed that the direct effect of the environment on the phenotype of a plant is generally in a similar direction to that in genetically controlled ecotypes, he found himself plagued by critics who accused him of having Lamarckian views. He remained opposed to the suggestion that this was evidence of the inheritance of acquired characters. Generally we accept his view. But this is not to believe that there is no evidence of direct influences of the environment on factors which are carried in the seed to the next generation or even several generations. Maternal effects are to be seen, particularly with seed and seedling characters, because the seeds are nourished by and surrounded by maternal tissues. And this influence is inflicted during the sensitive time of seed development and maturation.

We may look at a couple of examples: one on the range management scale and the other in the laboratory with very careful environmental and genetical control.

Taeniatherum asperum

Thus, take the investigation by Nelson, *et al.* (1970) on the Medusa head grass (*Taeniatherum asperum*). This annual species has spread rapidly through the grasslands of Washington, Oregon, and northern California after introduction from Eurasia. Taking plants at one location, they grew them at two stations which differed in humidity during the time of embryogenesis and maturation of the seeds. They found a significant increase in the rate of growth of plants raised from seeds that had been developed under conditions of greater humidity.

Abutilon theophrasti

Contrasting with this large-scale experiment is the study by Wulff & Bazzaz (1992) on an Old-World annual, *Abutilon theophrasti*, which has become a weed in parts of North America. Their experiment was carried out on ramets derived from only two genets. They exposed the experimental ramets to differing concentrations of Hoagland's solution at the time of their seed development. While sowing the seeds derived from the experiments and the control, they found increased seed weight as a result of the treatments. In the early growth stages of the seedlings derived from these seeds, the increase in the maternal nutrient supply increased seedling height, cotyledon area and leaf area, apparently deriving from the increased seed weight. After this came a period in which the maternal nutrient supply ceased to have an effect; but even after 66 days it did affect the seed and leaf dry weight of the second generation.

"The present results show that in *Abutilon theophrasti* a change in the parental nutrient supply may affect the growth of the progeny in a nutrient-variable environment, that there is variability in the response in the progeny of the two genotypes, and that parental nutrient addition may have a long-lasting effect on some growth characters of the progeny. This type of maternal effect may thus potentially have ecological and evolutionary significance. However, these experiments were conducted in controlled conditions where the possibilities of finding significant differences are high. The next question to be asked is to what extent these responses are important in a variable field environment." (Wulff & Bazzaz, 1992:107).

Maternal effects of all kinds on plants are reviewed by Roach & Wulff (1987).

Edaphic Factors

Although they may be more variable locally than are climatic factors, edaphic factors cannot be ignored in the ecotypical differentiation of weeds.

Most attention has been given to soils derived from ultramafic rocks — so-called serpentine soils. Reactions to serpentine soils have been studied mostly with native species following the first demonstration of serpentine-tolerant genotypes in California native species by Arthur Kruckeberg (1951) and the cooperation between Kruckeberg and Richard Walker. Kruckeberg, in his chapter in this book, covers the situation demonstratively but again there seems to be in the literature a relative lack of information about weedy species in relation to serpentine soils. Perhaps the situation is different in the weeds, and I suspect that experiments will show a tendency for GPG strategies to prevail in weeds.

In a doctoral thesis, Jean Whatley (1974) found no evidence of serpentine races in weed *Rumex crispus* growing on and off serpentine outcrops in the San Francisco Bay Area.

Although it is not a weed in the strict sense, *Pinus sabiniana is* at least a primary colonizer and is found, in California, on serpentine soils and non-serpentine soils. Seeds of this species collected from both kinds of habitat were grown by James Griffin and measurements of shoot and root growth were made. No adaptation to the serpentine soil conditions by populations which came from serpentine habitats was suggested. All seedling populations appeared to have high tolerance for extreme serpentine conditions (Griffin, 1965). This suggests a general purpose genotype (GPG) condition.

I would also remind you of the catholicity of soil variation tolerated by *Cortaderia jubata* (see page 198).

A similar concentration on native plants has taken place in studies of heavy metal tolerance. The literature on heavy metal tolerance is well reviewed by Antonovics, *et al.* (1971) and by Shaw (1989). It would be interesting to know if there is any latent tendency for the GPG strategy to apply here in the case of weeds that also invade these metalliferous soils. We should remember *Typha latifolia* described by McNaughton, *et al.* (1974; see page 202).

Not all weeds have wide soil tolerance. Thus, in California, escaped *Digitalis purpurea* is restricted to acid soils (as it is in its home territory in Europe) (Baker, 1974). It remains to be seen whether intraspecific variation in tolerance of acid and alkaline conditions is shown by weed populations as it is in the forage plants that have been investigated (*e.g.*, Snaydon & Davies, 1972). I predict that it will be less obvious in the weeds.

Resistance to Herbicides

Important for weeds is their susceptibility to damage by herbicidal sprays and the selection of biotypes that are resistant to the sprays is being investigated by numerous workers.

In a comprehensive review, Suzanne Warwick (1991) notes that at the time of writing over a hundred weed species were known to develop resistance to herbicides, with the first resistance found in *Senecio vulgaris* in 1968 (Ryan, 1970). Most of these cases of the development of resistance are to the triazine group of chemicals, but cases of development of resistance to other herbicides are substantiated and there are even some examples of resistance to more than one type of herbicide.

This research area is bound to have importance for the analogies that can be drawn between this selection process and others (like resistance to toxic soils) that are found in the microevolution of weeds. GPGs might also be found.

Demography

There is much need for studies by demographic means of the spread of weeds after introduction into a new area. Herbarium collections, photographs, and even seed catalogues can contribute to the study of the history of invasions (Mack, 1991).

For example, the spread of the autogamous annual grass, *Bromus tectorum,* has been studied very carefully by Richard Mack in the region of the steppe communities between the Rocky Mountains and the Cascade- Sierra Nevada Ranges. *Bromus tectorum* comes from arid Eurasia and, like so many of these species, was introduced by European farmers inadvertently or deliberately. The populations that had been established from introduced seed in the late 19th century appear to have been, for a period, in what Mack calls the 'lag phase.' The explosive expansion of the range of this species (after the lag phase) began about 1900 A.D. Some time after 1910, the rate of new range occupation was apparently limited only by the rate of production of seeds, but by 1930 the range expansion was virtually over (Mack, 1981, 1985, 1986; Mack & Pyke, 1983). The general studies and conclusions drawn by Mack for weed invasions are commendable to us all as the careful utilization of demographic techniques in history.

Other examples of demographic studies of weed spread are given in Mooney, *et al.* (1986), Baker (1986, 1989), and Mack (1986). But a famous case of the birth and spread of weedy material is surprisingly little discussed in the literature.

This concerns the so-called Oxford ragwort, *Senecio squalidus* L. This species, native on the Sicilian and southern Italian volcanoes was introduced to the Oxford University Botanic Garden before 1794 but was confined to old walls in the city of Oxford for very many years (Turrill, 1948) and very scattered occurences elsewhere (Salisbury, 1961). With the building of the railway between Oxford and London, it spread slowly along the chipped rock track ballast towards the capital city which it reached in time to experience the firebombing and explosive impact of the air raids of World War II. Suddenly *Senecio squalidus* became one of the commonest weeds in Great Britain. Although the plants are self-incompatible, Turrill (1948) reported that weedy London populations of *Senecio squalidus* are morphologically less variable than populations growing in the original area in Oxford, suggesting that not all genotypes were needed or even useful in the explosive spread of this weed, following its lengthy 'lag phase.' There is much that could be learned from the populations of this species which has also hybridized with other species of *Senecio* in Britain.

If a newly introduced plant does not have appropriate 'general purpose genotypes' available, it may be confined to a restricted area until these do become available through recombination or introgression. This may be one of the bases of the 'lag' phase so often seen in the history of a new weed, although many other factors may also contribute to this delay in spread, such as the necessity to build up an 'infection pressure' of propagules or the relaxation of a particular environmental deterrent (Salisbury, 1961).

Hybridization

A weed migrating across a geographical area may encounter native or introduced material of a related taxon. The taxa may be differentiated sufficiently for the weed to be unable to cross with the encountered party (*e.g.*, *Erodium moschatum* and *E. cicutarium* — both introduced from Europe into western American grasslands where they show great variation in environmentally sensitive characters but apparently never crossing). In Turesson's terminology they would be said to belong to different coenospecies.

If hybridization is possible, there may be one of three outcomes. Examples will be given of each kind of possible outcome.

Introgression

If there is a more or less fertile hybrid, there may be opportunities for production of further hybrid generations (probably mostly by back-crossing to the locally more abundant species). By this means infiltration of genes from one population into the other can take place. It may be the migrating taxon which will take up most of the 'new' genes. Heiser (1965) postulated that *Helianthus annuus* migrating westwards in North America picked up genes from local populations of other *Helianthus* species, and improved its local adaptation (however, see Rieseberg, *et al.*, 1988). More often it seems that the resident population is the one most obviously displaying the subtle influence of the hybridization.

Raphanus

An excellent example of introgression involving weedy species is to be seen in the radish genus, *Raphanus* (Brassicaceae). Introgressive hybridization has played an important role in the success of an erstwhile crop plant (*R. sativus*) as a weed in contemporary coastal California — in which it is climatically well-suited (Panetsos & Baker, 1968). All of the characteristics for which it has been selected as a crop plant (in Europe since Roman times) are against its success as a weed: rapid, simultaneous germination, persistence in a rosette instead of bolting into flower, development of a swollen non-deeply penetrating root system, and non-dispersing but soft-walled fruit.

Raphanus raphanistrum (wild radish), introduced into California as a grain field and pasture weed (also from Europe), can provide (and has supplied by hybridization) the necessary characteristics for weed growth. These include discontinuous germination, bolting from the transient rosette, development of a penetrating branched root system, hard(brittle)-walled fruit that is dispersed as single- seeded segments. It does not seem to be as well-adapted to coastal environments as *R. sativus* (Baker, 1974) and its purer populations are in the foothills of the Sierra Nevada and the eastern parts of the Central Valley.

Artificial hybrids between *R. raphanistrum* and a cultivated form of *R. sativus* exhibited about 50% pollen fertility and were heterozygous for a rather large reciprocal translocation, which produced a ring of four chromosomes (plus ten bivalents) at meiosis. "Examination of 'wild' populations of *R. sativus* revealed that plants heterozygous for the reciprocal translocation are present in varying proportions.

Thus a cytological proof of the introgression is added to the morphological evidence" (Panetsos & Baker, 1968:243).

Raphanus in California is a hybrid swarm and this should be kept in mind when using radish material for evolutionary experiments which have been becoming increasingly frequent in the literature.

A very recent review of "Natural Hybridization as an Evolutionary Process" by Michael Arnold (1992) substantiates introgression as an evolutionary force by the utilization of molecular biology methods and reviews the contributions of gene-flow through the pollen and seed DNA constitutions.

Alternatively to the formation of fertile hybrids, the outcome of crossing two species may be a sterile hybrid which may double its chromosome number and become a fertile allopolyploid.

Hybridization of Distantly Related Species

Tragopogon

An example of this process is given by the brilliant investigation carried out by Marion Ownbey in the vicinity of Moscow, Idaho (Ownbey, 1950). Here there occur, on roadsides and in disturbed places, three diploid species of introduced (outcrossing) *Tragopogon* (*T. porrifolius, T. dubius*, and *T. pratensis*) (Asteraceae).

These Eurasian species form sterile hybrids but, on a background of sterile hybrid plants, Ownbey found groups of fertile plants. These proved to be of two different tetraploids (amphidiploid of *T. dubius* × *T. porrifolius* and of *T. dubius* × *T. pratensis*). These tetraploids were named as new species, *T. mirus* and *T. miscellus,* respectively (Ownbey, 1950). The spread of these new species is being monitored. *Tragopogon miscellus* has already become a common weed in the vicinity of Spokane, Washington (Roose & Gottlieb, 1976).

Spartina

An environmentally important case of hybridization between an invader and a resident species, resulting in an allopolyploid species' birth and distribution, is to be seen in the cord grass genus *Spartina*. *Spartina alterniflora* (2n = 62), native to salt marshes of the East Coast of North America, was introduced fortuitously into a salt marsh in Southampton Water (on the south coast of England), where it met the resident *S. maritima* (also diploid, 2n = 60). Hybridization occurred and the vigorous diploid hybrid (*S.* × *townsendii*, 2n = 62) was formed. This spread widely on the European coast, forming almost pure stands in salt marshes, but the diploid hybrid is now accompanied by a tetraploid (amphididiploid) derivative — *S. anglica* (4n = 120-124) — a new species (Clapham, *et al.*, 1987; Thompson, 1991). The diploid was able to colonize extensively without seed formation, but the tetraploid has the extra advantage of seed production. These derivatives of *Spartina* are of extreme economic importance, and have been widely planted to stabilize salt marshes, although they crowd out other species of salt marsh plants. The pure stands of *Spartina* provide an excellent center for pathogens and already ergot (*Claviceps purpurea*) has made serious impacts locally (Thompson, 1991). Presumably selection for resistance to these pathogens will now take place.

EPILOGUE

It is hoped that some of the needs in the expansion of our understanding of the genecology of weeds will be apparent, with the recognition that the principles of ecotypical differentiation can be studied with human-influenced plants as well as with natives.

But we must take account of the rise of molecular biology and realize the complexity of genetic systems that depend upon active DNA in the nucleus, chloroplasts, and mitochondria. We must take into account the evidence of redundant multiplication of individual genes that may remain 'silent' or be energized by 'regulatory genes' some of which may be 'transposable genetic elements' (Barbara McClintock's 'jumping genes') which can excise themselves from a chromosome and attach themselves at other places in the genome where they may have a regulatory function. These influences may have some effect on the stability of the genotype and therefore have genecological consequences. We do not know what proportion of the genome in any taxon is made up of these elements.

It is my belief that the results of experiments carried out under the mandate of *pre-molecular* biology will tend to be still usable in the face of the revelations of *molecular* biology. It all depends on the level at which one is working.

The molecular approach (and the more physiological aspects of genecology) requires 'high-tech' laboratory and growing facilities and highly trained technicians. This would seem to leave individuals and groups with limited means out of the picture, but it need not be so, as long as the limitations of the observations and experiments that can be meaningful are realized. There is still a lot of substantiation and filling-in of the genecological picture to be done and careful observation followed by simple experimentation is within the bounds of possibility for amateur biologists (Baker, 1982).

Just as we can see good biology in the work of Göte Turesson and his followers in the past eighty years, so there will be a modicum of importance in every careful genecological study that is made. For these less ambitious investigations, weeds may not only be more suitable but more easily available than the native plants of any area (whose conservation may have to take precedence over their utilization in experiments).

ACKNOWLEDGEMENTS

I proffer my thanks to Drs. A. R. Kruckeberg and R. B. Walker for the invitation to contribute to this symposium. This would have been impossible without the labors of my dedicated amanuenses Johanna Jones and Kim Nguyen-Tan. My debt to all my colleagues who have guided me through many years of my concern with weeds is apparent.

References

Allard, R. W., G. R. Babble, M. T. Clegg, and A. L. Kahler. 1972. Evidence for coadaptation in *Avena barbata*. *Proc. Nat'l. Acad. Sci. USA* 69:3043-48.

Antonovics, J., and A. D. Bradshaw. 1970. Evolution in closely adjacent plant populations. VIII. Clinal patterns at a mine boundary. *Heredity* 25:349-362.

Antonovics, J., A. D. Bradshaw, and R. G. Turner. 1971. Heavy metal tolerance in plants. *Adv. Ecol. Res.* 7:1-85.

Arnold, M. L. 1992. Natural hybridization as an evolutionary process. *Annu. Rev. Ecol. Syst.* 23: 237-262.

Baker, H. G. 1953. Race formation and reproductive method in flowering plants. *Symp. Soc. Exp. Biol.* 7:114-145.

Baker, H. G. 1954. Report of paper presented to the British Ecological Society. *Jour. Ecol.* 42:571.

Baker, H. G. 1955. Self-compatibility and establishment after 'long-distance' dispersal. *Evolution* 9:347-348.

Baker, H. G. 1965. Characteristics and modes of origin of weeds.

Pages 147-172 *in* H.G. Baker & G. L. Stebbins, eds., *The Genetics of Colonizing Species*. Academic Press, New York.

Baker, H. G. 1966. The evolution, functioning and breakdown of heteromorphic incompatibility systems. 1. The Plumbaginaceae. *Evolution* 20:349-368.

Baker, H. G. 1972a. Human influences on plant evolution. *Econ. Bot.* 26:32-43.

Baker, H. G. 1972b. Migrations of weeds. Pages 327-347 *in* D. H. Valentine, ed., *Taxonomy, Phytogeography and Evolution*. Academic Press, London.

Baker, H. G. 1974. The evolution of weeds. *Annu. Rev. Ecol. Syst.* 5:1-24.

Baker, H. G. 1978. Invasion and replacement in Californian and neotropical grasslands. Chapter 24, pages 368-384 *in* J. R. Wilson, ed., *Plant Relations in Pastures*. C.S.I.R.O., East Melbourne, Australia.

Baker, H. G. 1982. A tribute to the amateur in Botany. *Univ. Washington Arboretum Bull.* 45(3):10-17.

Baker, H. G. 1985. What is a weed? *Fremontia* 12:7-11.

Baker, H. G. 1986. Patterns of plant invasion in North America. Chapter 3, pages 44-57 *in* H. A. Mooney & J. A. Drake, eds., *Ecology of Biological Invasions of North America and Hawaii*. Springer-Verlag, New York.

Baker, H. G. 1989. Sources of the naturalized grasses and herbs in Californian grasslands. Pages 29-38 *in* L. F. Huenneke & H. A. Mooney, eds., *Grassland Structure and Function. California Annual Grassland.* Kluwer Academic Publishers, Dordrecht, Netherlands.

Baker, H. G. 1991. The continuing evolution of weeds. *Econ. Bot.* 45:445-449.

Baker, H. G. and P. A. Cox. 1984. Further thoughts on dioecism on islands. *Ann. Missouri Bot. Gard.* 71:230-239.

Barrett, S. C. H. 1980. Sexual reproduction in *Eichhornia crassipes* (water hyacinth), II Seed production in natural populations. *Jour. Appl. Ecol.* 17:113-124.

Barrett, S. C. H. 1982. Genetic Variation in Weeds. Pages 73-98 *in* R. Charudattan & H. Walker, eds., *Biological Control of Weeds by Pathogens*. Wiley, New York.

Barrett, S. C. H. 1983. Crop mimicry in weeds. *Econ. Bot.* 37:255-282.

Barrett, S. C. H. 1988. Genetics and evolution of agricultural weeds. Pages 73-98 *in* M. Altieri & M. Liebman, eds., *Weed Management in Agroecosystems: Ecological Approaches*. CRC Press, Inc., Boca Raton, FL.

Barrett, S. C. H. 1992. Genetics of weed invasions. Pages 91-119 *in* S. K. Jain & L. W. Botsford, eds., *Applied Population Biology*. Kluwer Academic Publishers, Dordrecht, Netherlands.

Barrett, S. C. H. and B. C. Husband. 1989. The genetics of plant migration and colonization. Pages 254-277 *in* A. H. D. Brown, M. T. Clegg, A. L. Kahler, & B. S. Weir, eds., *Plant Population Genetics, Breeding, and Genetic Resources*. Sinauer, Sunderland.

Barrett, S. C. H. and D. E. Seaman. 1980. The weed flora of Californian rice fields. *Aquat. Bot.* 9:351-376.

Barrett, S. C. H. and J. S. Shore. 1988. Isozyme variation in colonizing plants. Pages 106-126 *in* D. E. Soltis & P. S. Soltis, eds., *Isozymes in Plant Biology*. Dioscorides Press, Washington.

Bosbach, K. and H. Hurka. 1981. Biosystematic studies of *Capsella bursa-pastoris* (Brassicaceae): Enzyme polymorphism in natural populations. *Plant Syst. Evol.* 137:73-94.

Bradshaw, A. D. 1959. Population differentiation in *Agrostis tenuis* Sibth. I. Morphological differentiation. *New Phytol.* 58:205-227.

Bradshaw, A. D. 1960. Population differentiation in *Agrostis tenuis* Sibth. III. Populations in varied environments. *New Phytol.* 59:92-103.

Bradshaw, A. D. 1965. Phenotypic plasticity in plants. *Brookhaven Symp. Biol.* 25:75-94.

Bradshaw, A. D. and R. W. Snaydon. 1959. Population differentiation within plant species in response to soil factors. *Nature* 183:129-130.

Brown, A. H. D., and D. R. Marshall. 1981. Evolutionary changes accompanying colonization in plants. Pages 351-363 *in* G. G. E. Scudder & J. K. Reveal, eds., *Evolution Today*. Proc. Second Internat. Congr. Systematic and Evolutionary Biol. Carnegie-Mellon Univ., Pittsburgh, Pennsylvania.

Chapin, F. S., III, K. Autumn, and F. Pugnaire. 1993. Evolution of suites of traits in response to environmental stress. *American Nat.* 142:S78-S92.

Clapham, R. C., T. G. Tutin and D. M. Moore. 1987. *Flora of the British Isles*, 3rd ed. Cambridge University Press, Cambridge.

Clausen, J. 1949. Genetics of climatic races of *Potentilla glandulosa. Proc. Eighth Internat. Congr. Genetics, Hereditas* (suppl.vol) :162-172.

Clausen, J. 1953. New bluegrasses by combining and rearranging genomes of contrasting *Poa* species. *Proc. 6th Internat. Grassland Congr.*, University Park, Pennslyvania, 1952, pp. 216-221.

Clausen, J., D. D. Keck, and W.M. Hiesey. 1939. The concept of species based on experiment. *American Jour Bot.* 26:103-106.

Clausen, J., D. D. Keck, and W. M. Hiesey. 1940. Experimental Studies on the Nature of Species. I. Effect of varied environments on western North American plants. *Carnegie Inst. Washington, Publ.* 520. 450 pp.

Connor, H. E. 1973. Breeding systems in *Cortaderia* (Gramineae). *Evolution* 27:663-678.

Connor, H. E. and E. Edgar. 1974. Names and types in *Cortaderia* Stapf. (Gramineae). *Taxon* 23:595-605.

Costas-Lippmann, M. A. 1976. Ecology and Reproductive Biology of the Genus *Cortaderia* in California. *Ph.D. thesis.* Department of Botany, University of California, Berkeley.

Crawley, M. J. 1986. The population biology of invaders. *Philos. Trans. Roy. Soc. London*, Ser. B, 314:711-731.

Geber, M. A., M. A. Watson, and R. Furnish. 1992. Genetic differences in clonal demography in *Eichhornia crassipes. Jour. Ecol.* 80:329-342.

Godwin, H. 1960. The history of weeds in Britain. Pages 1-10 *in* J. L. Harper, ed., *The Biology of Weeds*. Blackwell's, Oxford, UK.

Griffin, J. R. 1965. Digger pine seeding response to serpentinite and non-serpentinite soil. *Ecology* 46:801-807.

Grime, J. P. 1977. Evidence for the existence of three primary strategies in plants and its relevance to ecological and evolutionary theory. *American Nat.* 111:1169-1194.

Hamrick, J. L. and R. W. Allard. 1972. Microgeographical variation in allozyme frequencies in *Avena barbata. Proc. Natl. Acad. Sci, USA* 65:2100-4.

Hamrick, J. L., Y. B. Linhart, and J. B. Mitton. 1979. Relationships between life history characteristics and electrophoretically detectable genetic variation in plants. *Annu. Rev. Ecol. Syst.* 10:173-200.

Harper, J. L. (ed.). 1960. *Biology of Weeds*. British Ecol. Soc. Symp. No. 1. Blackwell's, Oxford, UK.

Hegi, G. 1929. *Illustrierte Flora von Mittel-Europa.* VI(2):584-587. J.F. Lehmans Verlag, München.

Heiser, C. B. Jr. 1965. Sunflowers, weeds and cultivated plants. Pages 391-401 *in* H. G. Baker & G. L. Stebbins, eds., *The Genetics of Colonizing Species*. Academic Press, London.

Heslop-Harrison, J. 1964. Forty years of genecology. *Adv. Ecol. Res.* 2:159-247.

Holmgren, I. 1919. Zytologische Studien über die Fortpflanzung bei den Gattungen *Erigeron* und *Eupatorium. Kgl. Svenska Vetenskapsakad. Handl.* 59(7):1-117.

Hume, L. and P. B. Cavers. 1981. A methodological problem in genecology. Seeds versus clones as source material for uniform gardens. *Canadian Jour. Bot.* 59:763-768.

Jennings, D. H., and A. L. Trewavas (eds.). 1986. *Plasticity in Plants.* U.K. Company of Biologists, Cambridge.

Kruckeberg, A. R. 1951. Intraspecific variability in the response of certain native plant species to serpentine soil. *American Jour. Bot.* 38:408-419.

Lefèbvre, C. 1970. Self-fertility in maritime and zinc mine populations of *Armeria maritima* (Mill.) Willd. *Evolution* 24:571-577.

Lefèbvre, C. and Vernet, P. 1976. Microevolutionary processes on contaminated deposits. Pages 285-300 *in* J. Shaw, ed., *Heavy Metal Tolerance in Plants: Evolutionary Aspects*. CRC Press, Boca Raton, Florida.

Mack, R. N. 1981. Invasion of *Bromus tectorum* L. into western North America: An ecological chronicle. *Agro-Ecosytems* 7:145-165.

Mack, R. N. 1985. Invading plants: Their potential contribution to population biology. Pages 127-142 *in* J. White, ed., *Studies on Plant Demography*. John L. Harper Festschrift. Academic Press, London.

Mack, R. N. 1986. Alien plant invasion into the Intermountain West: a case history. Pages 191-213 *in* H. A. Mooney & J. A. Drake, eds., *Ecology of Biological Invasions of North America and Hawaii*. Ecological Studies 58. Springer-Verlag, New York.

Mack, R. N. 1991. The commercial seed trade: An early dispenser of weeds in the United States. *Econ. Bot.* 45:257-273.

Mack, R. N. and D. A. Pyke. 1983. The demography of *Bromus tectorum*: Variation in time and space. *Jour. Ecol.* 71:69-93.

McNaughton, S. J. 1966. Ecotype function in the *Typha* community-type. *Ecol. Monogr.* 36:296-325.

McNaughton, S. J. 1970. Fitness sets for *Typha. American Nat.* 104:337-342.

McNaughton, S. J. 1974. Development control of net productivity in *Typha latifolia* ecotypes. *Ecology* 55:168-172.

McNaughton, S. J., T. C. Folsom, T. Lee, F. Park, C. Prive, D. Roeder, J. Schmits, and C. Stockwell. 1974. Heavy metal tolerance in *Typha latifolia* without the evolution of tolerant races. *Ecology* 55:1163-1165.

Moe, L. M. 1977. Ecological and Systematic Studies on the Discoid *Matricarias* of North America. *Ph.D. Thesis* (Botany). Univ. California, Berkeley.

Mooney, H. A. and J. A. Drake (eds.). 1986. *Ecology of Biological Invasions of North America and Hawaii.* Springer-Verlag, New York.

Mortimer, A. M. 1983. On weed demography. Pages 3-41 *in* W. W. Fletcher, ed., R*ecent Advances in Weed Research.* Commonwealth Agric. Bur., Farnham Royal, England.

Munz, P. A. 1963. *A California Flora.* Univ. California Press, Berkeley.

Nelson, J. R., G. A. Harris, and C. J. Goebel. 1970. Genetic vs. environmentally induced variation in medusa head (T*aeniatherum asperum* (Simonkai) Nevski). *Ecology* 51:526-529.

Ownbey, M. 1950. Natural hybridization and amphiploidy in the genus *Tragopogon. American Jour. Bot.* 37:487-499.

Panetsos, C. A. and H. G. Baker. 1968. The origin of variation in 'wild' *Raphanus sativus* (Cruciferae) in California. *Genetica* 38:243-274.

Pritchard, T. 1960. Race formation in weedy species with special reference to *Euphorbia cyparissias* L. and *Hypericum perforatum* L. Pages 61-66 *in* J. L. Harper, ed., *Biology of Weeds.* Cambridge Univ. Press, Cambridge.

Quinn, J. A., and J. C. Colosi. 1977. Separating genotype from environment in germination ecology studies. *American Midl. Nat.* 97:484-489.

Rapson, G. L. and J. B. Wilson. 1988. Nonadaptation in *Agrostis Capillaris* L. (Poaceae). *Functional Ecol.* 2:479-490.

Rapson, G. L. and J. B. Wilson. 1992a. Genecology of *Agrostis capillaris* L. (Poaceae) – an invader into New Zealand.1. Floral phenology. *New Zealand Jour. Bot.* 30:1-11.

Rapson, G. L. and J. B. Wilson. 1992b. Genecology of *Agrostis capillaris* L. (Poaceae) – an invader into New Zealand. 2. Responses to light, soil fertility, and water availability. *New Zealand Jour. Bot.* 30:13-24.

Rejmanek, M., C. D. Thomsen and J. D. Peters. 1991. Invasive vascular plants of California. Chapter 8, pages 31-101 *in* R. H. Groves & F. diCastri, eds., *Biography of Mediterranean Invasions*. Cambridge Univ. Press, London.

Rieseberg, L. H., D. E. Soltis, and J. D. Palmer. 1988. A molecular reexamination of introgression between *Helianthus annus* and *H. bolanderi* (Compositae). *Evolution* 42:227-238.

Richards, A. J., C. Lefèbvre, M. G. Macklin, A. Nicholson, and X. Vekemans. 1989. The population genetics of *Armeria maritima* (Mill.) Wild. on the River South Tyne, UK. *New Phytol.* 112:281-293.

Roach, D. A. and R. D. Wulff. 1987. Maternal effects in plants. *Annu. Rev. Ecol. Syst.* 18:209-235.

Roose, M. L. and L. D. Gottlieb. 1976. Genetic and biochemical consequences of polyploidy in *Tragopogon*. *Evolution* 30:818-830.

Ryan, G. F. 1970. The resistance of common grounsel to simaizine. *Weed Sci.* 34:40-48.

Salisbury, E. J. 1961. *Weeds and Aliens*. Collins, London.

Sarukhan, J. 1974. Studies on plant demography in *Ranunculus repens* L., *R. bulbosus* L., and *R. acris* L. II. Reproductive strategies and seed population dynamics. *Jour. Ecol.* 62:151-177.

Shaw, J. (ed.). 1989. *Heavy Metal Tolerance in Plants: Evolutionary Aspects*. CRC Press, Boca Raton, Florida.

Shull, G. H. 1937. The geographical distribution of the diploid and double-diploid species of shepherds' purse. *Nelson Fithian Birthday Volume*, pages 1-8. Published privately in Boston under chairmanship of Heber K. Youngken.

Snaydon, R. W., and M. S. Davies. 1972. Rapid population differentiation in a mosaic environment. I. The response of *Anthoxanthum odoratum* populations to soils. *Evolution* 24:257-269.

Stearns, S. C. 1989. The evolutionary significance of phenotypic plasticity. *BioScience* 39:436-445.

Stebbins, Jr. G. L. 1950. *Variation and Evolution in Plants*. Oxford Univ. Press, London.

Thompson, J. B. 1991. Biology of an invasive plant. What makes *Spartina anglica* so succesful? *BioScience* 41:393-401.

Turesson, G. 1922. The genotypical response of the plant species to the habitat. *Hereditas* (Lund) 3:211-350.

Turesson, G. 1925. The plant species in relation to habitat and climate. Contributions to the knowledge of genecological units. *Hereditas* (Lund) 6:147-236.

Turrill, W. B. 1948. *British Plant Life*. Collins, London.

Vekemans, X., C. Lefèbvre, L. Belallia, and P. Meerts. 1990. The evolution and breakdown of the heteromorphic incompatibility system of *Armeria maritima* revisited. *Evol. Trends in Plants* 4:15-24.

Warwick, S. I. 1991. Herbicide resistance in weedy plants: Physiology and population biology. *Annu. Rev. Ecol. Syst.* 22:95-114.

Warwick, S. I. and D. Briggs. 1978. The genecology of lawn weeds I. Population differentiation in *Poa annua* in a mosaic environment of bowling green lawns and flower beds. *New Phytol.* 81:711-728.

Whatley, J. 1974. Comparative Studies on some Components of Serpentine and Non-serpentine Vegetation. *Ph.D. Thesis* (Botany). Univ. California, Berkeley.

Wilson, F. 1960. *A Review of the Biological Control of Insects and Weeds in Australia and Australian New Guinea*. Tech. Comm. No. 1. Commonwealth Institute of Biological Control, Commonwealth Agricultural Bureaux. Farnham Royal, England.

Wulff, R. D. and F. A. Bazzaz. 1992. Effect of the parental nutrient regime on growth of the progeny in *Abutilon theophrasti* (Malvaceae). *American Jour. Bot.* 79:1102-1107.

Ecophenotypic Variation in the Territorial Songs of Galápagos Finches

Robert I. Bowman
Department of Biology
San Francisco State University
San Francisco, California 94118

Ornithologists have long been aware of regional variations in the vocal expressions of songbirds (partial summary in Mundinger, 1982). In recent years detailed sound spectrographic analyses of songs have provided objective documentation of this widespread "dialect phenomenon," which, in many cases, develops as a result of a filial tradition established by vocal learning. In the case of Galápagos finches, laboratory experiments with hand-reared individuals led to the conclusion that geographical vocal variants appear not to have a genetic basis (Bowman, 1983; Millington & Price, 1985), and therefore are best described as ecogeographic phenotypes or ecophenotypes (Mayr, 1963).

In recent years studies on habitat acoustics have indicated that sound transmission peculiarities of environments significantly affect the form of long-distance communication signals (Konishi, 1970; Chappuis, 1971; Jilka & Leisler, 1974; Morton, 1975; Marten & Marler, 1977; Marten, Quine & Marler, 1977; Bowman, 1979; Hunter & Krebs, 1979; Wiley & Richards, 1982; Ryan & Sullivan, 1989; and Ryan & Wilczynski, 1991). Abundant reports of inter- and intra-island variations in the advertising songs of Galápagos finches (Rothschild & Hartert, 1899, 1902; Snodgrass & Heller, 1904; Gifford, 1919; Beebe, 1924, 1926; and Lack, 1945) prompted Bowman to undertake major field and laboratory studies based on the prediction that, in the main, habitat type influences song structure, resulting in regional vocal dialect formation.

In this report I summarize data that correlate vocal dialects in Galápagos finches with sound attenuation characteristics of the natural environments, and with the finch biology. It is proposed that most of the adaptive properties of the songs are shaped by natural selection, including specific song dialects on specific islands, parallel shifts in songs of sympatric species, and intra-island variation in the songs of one species living in different habitats.

METHODS

A general description of the equipment and procedures used in the analysis of songs and environmental sound transmission follows. Details are presented elsewhere (see Bowman, 1983).

Field Recordings. — All songs were recorded on single track Nagra tape recorders (models III-B and IV) operating at 7.5 or 15 ips. Early in the study American D-33 microphones were mounted in 24-inch unpolished aluminum parabolic reflectors, but these were soon replaced with Sennheiser ultradirectional microphones, (models MKH 805 and 815U) equipped with windscreens.

Laboratory analysis of songs. — Recorded songs were fed into a Kay Elemetrics Spectrum Analyzer ("Sonagraph" model 6061B) equipped with an amplitude display from which energy spectra histograms were constructed for individuals and populations (Fig. 1).

From the song energy spectra the relative amplitude distribution (in percent) was determined, noting in particular the frequencies at which peak energy occurs (*i.e.*, "modal frequency") and also the upper and lower frequencies that delimit a bandwidth containing approximately 80-85% of the sound energy of a song population. This bandwidth is projected as a shaded band across the sound isopleth (see Fig. 2).

Measurement of decibel level of songs. — Field measurement of sound-pressure levels of territorial songs were made for five species of Galápagos finches, namely, *Geospiza magnirostris, G. conirostris, G. fuliginosa*, and *Certhidea olivacea* on Isla Genovesa; and *G. magnirostris, G. difficilis*, and *Certhidea* on Isla Wolf (Fig. 3).

A hand-held microphone equipped with a random-incidence corrector and windscreen was mounted on a microphone boom and connected to a sound level meter by means of a ten-foot long shielded extension cable. The microphone was calibrated with a pistonphone (B&K type 4220) and the non-weighted signal from the sound level meter was fed into a Nagra IIIB (*i.e.*, IIIB) tape recorder operating at 15 ips at the HIFI record setting, the two instruments have previously been "matched."

Measurement of sound transmission. — Sound transmission in

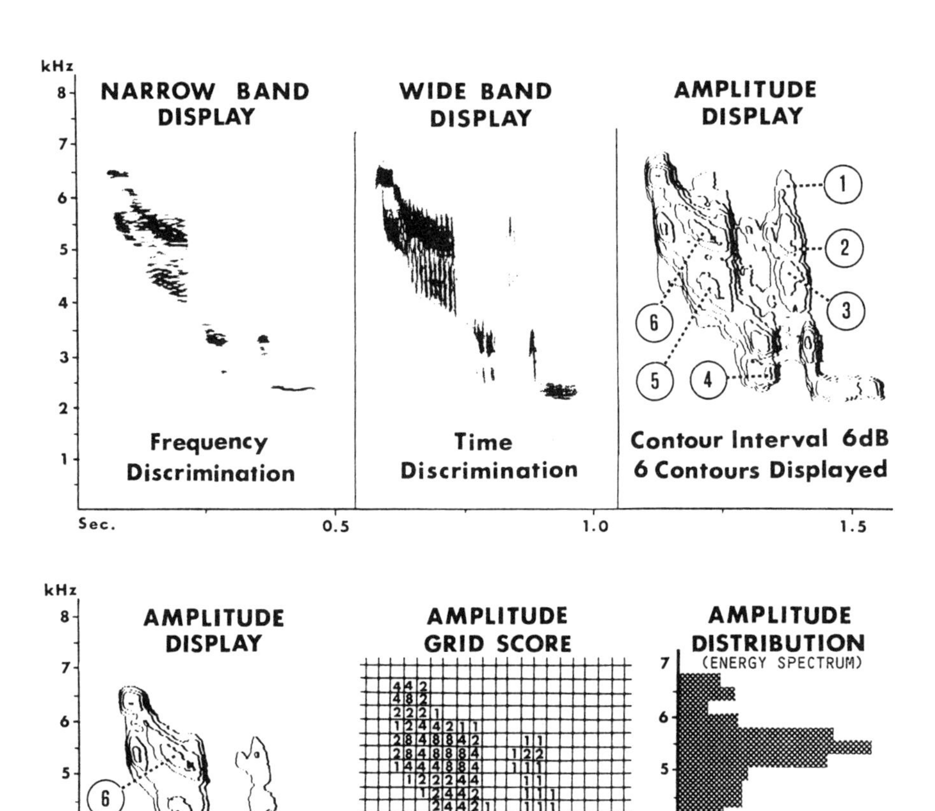

FIGURE 1. Comparison of three types of graphic display of bird song (above) and methods of analysis of relative amplitude of song (below). After Bowman, 1979.

song environments was measured on five islands in the archipelago, namely, Santa Cruz (Academy Bay transect from the outer coast to the inner coast to the lower transition to the upper transition to the *Scalesia* forest; Figs. 4a and 4b), Santiago (James Bay parkland), Española (two locations in the north coastal vegetation; Fig. 4c), Genovesa (north coastal vegetation of Darwin Bay; Fig. 4d), and Wolf (north shore plateau; Fig. 4e.)

Measurements were made early in the morning when the air was

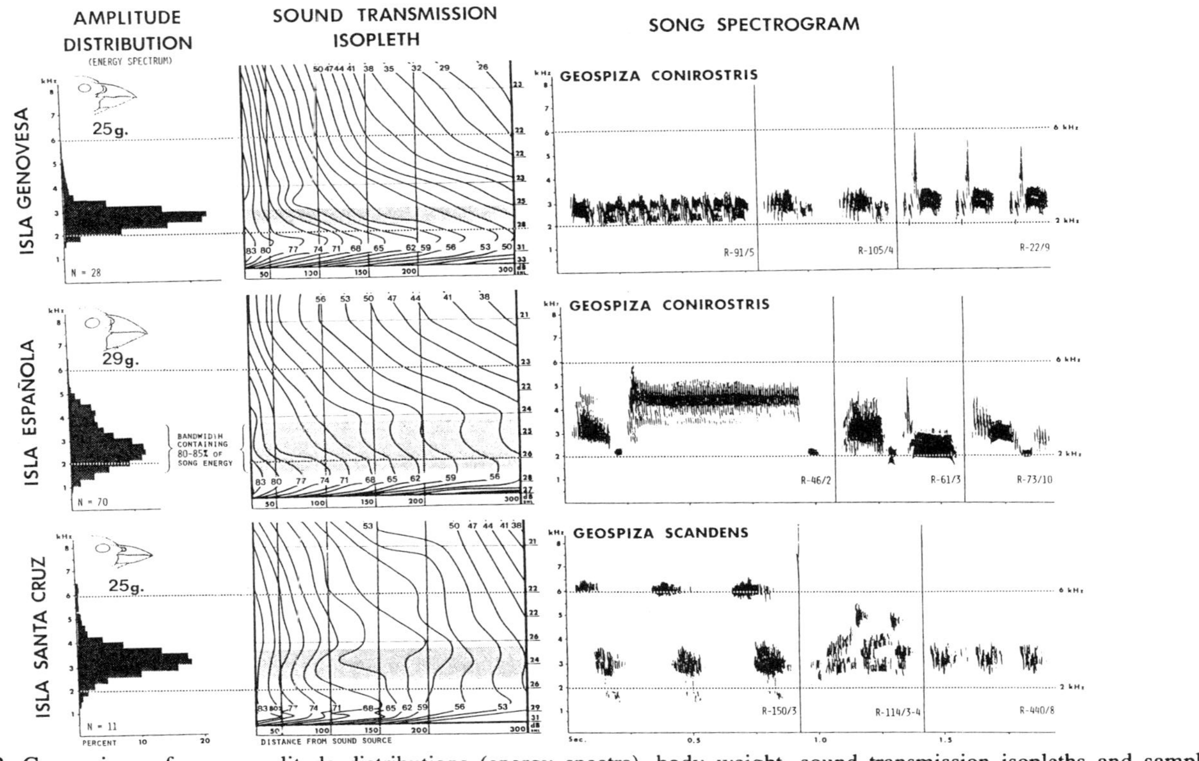

FIGURE 2. Comparison of song amplitude distributions (energy spectra), body weight, sound transmission isopleths and sample song spectrograms for *Geospiza conirostris* populations on Isla Genovesa (top), Isla Española (middle), and for *Geospiza scandens* on Isla Santa Cruz (bottom). Shaded bands on isopleths indicate bandwidths containing approximately 80-85% of the sound energy. Body weight datum for Isla Genovesa courtesy of P.R.Grant.

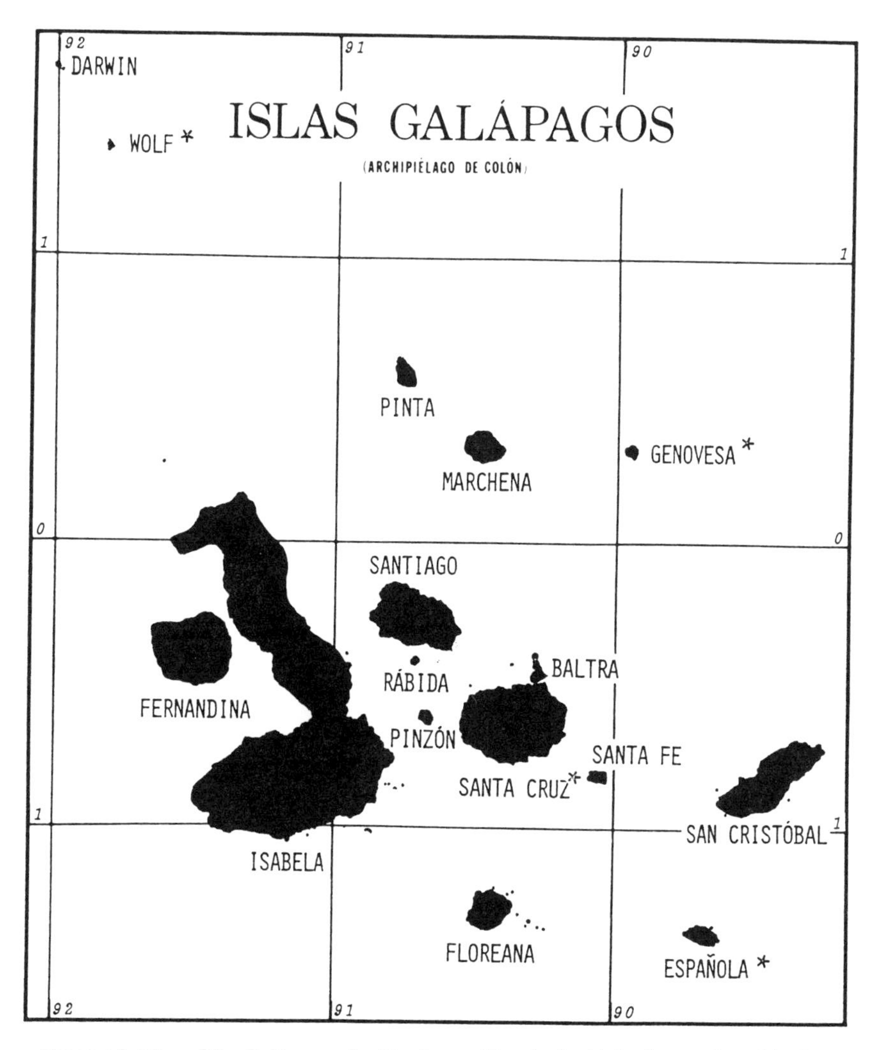

FIGURE 3. Map of the Galápagos Archipelago with principal islands mentioned in the text flagged with an asterisk (*).

calm and relatively free of heat stratification. An amplified "pink" noise spectrum, originating from a battery-powered noise generator, was broadcast through a speaker system mounted on a tripod 1 meter above ground level (Fig. 5). The transmitted noise was recorded through a sound level meter mounted on a tripod 1 meter above ground level and positioned within the vegetation at six stations along a straight line transect, 25, 50, 100, 150, 200 and 300 feet from the front of the speaker horn. The Nagra power amplifier was used as a

FIGURE 4a. Galápagos vegetation, Isla Santa Cruz. Above: Inner coastal zone, north shore of Academy Bay. Tree cacti are *Jasminocereus thousarsi* and *Opuntia echios*. Below: Transition zone, 150 m elevation, north of Academy Bay. Major trees are *Piscidia carthagenensis, Psidium galapageium*, and *Pisonia floribunda* (fallen trunk and large butressed tree at far right).

FIGURE 4b. Galápagos vegetation, Isla Santa Cruz. *Scalesia* forest zone, 500 m elevation, north of Academy Bay. The dominant tree is the composite *Scalesia pedunculata.* The understory is composed of the shrubs *Psychotria rufipes, Darwiniothamnus tenuifolius* and the tree *Zanthoxylum fagara,* among others.

portable power source for the hornspeaker. The fidelity of the power amplifier was high, having a uniform response from 20 Hz to 20,000 Hz. The frequency response of the sound level meter, measured between the microphone input and the output of the SLM, was perfectly flat between 20 and 20,000 Hz on the linear scale.

Recordings of transmitted pink noise were analyzed by means of a sound filter at one-third octave intervals from 500 Hz to 10,000 Hz. Sound pressure levels (dB values) were obtained from the strip chart of the graphic level recorder (Fig. 6).

Before filtering, the pink noise spectrum between 1 and 20 kHz was characterized by a sound pressure level (SPL) decrease of 3 dB per octave. After one-third octave filtering, the pink noise spectrum is converted to a white noise spectrum, characterized by a constant energy distribution per octave.

Tape-recorded pink noise signals were electronically separated into 14 frequency bands, *i.e.*, one-third octave bands with center frequencies at 500, 620, 790, 1000, 1240, 1580, 2000, 2480, 3160,

FIGURE 4c. Galápagos vegetation. Isla Española, Punta Suarez coastal region. Shrubs of *Alternanthera echinocephalus* and *Grabowskia boerhaaviaefolia* during the dry season (above) and wet season (below).

FIGURE 4d. Galápagos vegetation. Isla Genovesa, coastal region north of Darwin Bay. Dwarf *Bursera graveolens* forest during the wet season. Not shown is the dense growth of *Opuntia helleri,* most abundant along the shore where *Geospiza conirostris* is most abundant.

FIGURE 4e. Galápagos vegetation. Isla Wolf, northwest mesa, wet season. *Croton scouleri* thicket (left) and *Opuntia helleri* patch in foreground (right).

4000, 4960, 6320, 8000 and 99920 Hz. None of the songs of Galápagos finches contained fundamental frequencies below 500 Hz.

Display of sound transmission data. — One-third octave absolute sound-pressure levels (SPL's) have been plotted on graph paper having the frequency scale on the vertical axis and distance scale in feet from the sound source on the horizontal axis (Fig. 7). Points of equal absolute sound pressure are connected to produce an "isodecibel" (amplitude) contour line. In completed isopleths, contour lines are plotted at 3-dB intervals. The dB values for each contour line are shown near the top and bottom of each line. In this manner a "sound-transmission isopleth" is generated, showing the frequency-dependent pattern of sound attenuation in a given environment, i.e. the vegetation-induced acoustic "climate."

The energy spectra (frequency/amplitude histograms) of song populations may be "projected" onto the corresponding "song fields" (sound-transmission isopleths) to allow for easy visual detection of possible correlations between the most intensively used frequencies in the songs and the most energy-conserving transmission channels (frequency bandwidths) in a particular vegetative environment (Fig. 8).

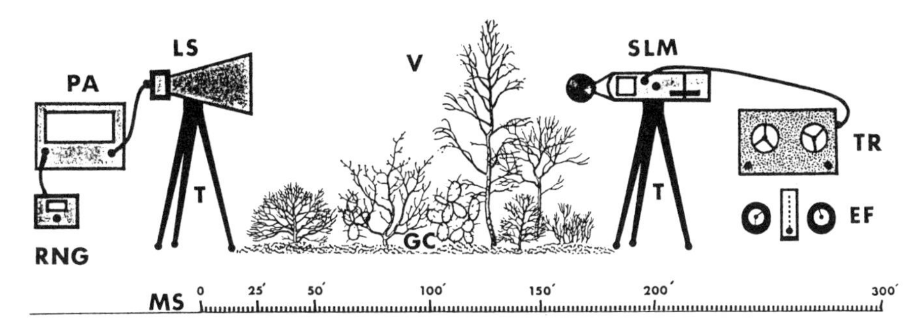

FIGURE 5. Arrangement of equipment used for measuring sound transmission in Galápagos environments. Key to symbols: EF, environmental factors (temperature, relative humidity, barometric pressure); GC ground conditions; LS loud speaker; MS monitoring stations; PA, power amplifier; RNG, random noise generator; SLM, sound level meter; T, tripod; IR, tape recorder; and V, vegetation.

In those Galápagos environments thus far tested, the most energy efficient channels for sound transmission occur in the frequency range of 1.5-2.0 kHz (the "sound window" of Morton, 1970, 1975). However, this is below the frequencies of modal amplitudes but within the lower range of most Galápagos finch songs.

From the sound-transmission isopleths we determine the pattern formed by the amplitude contour lines. At a specific frequency and distance from the sound source, where contour lines are narrowly

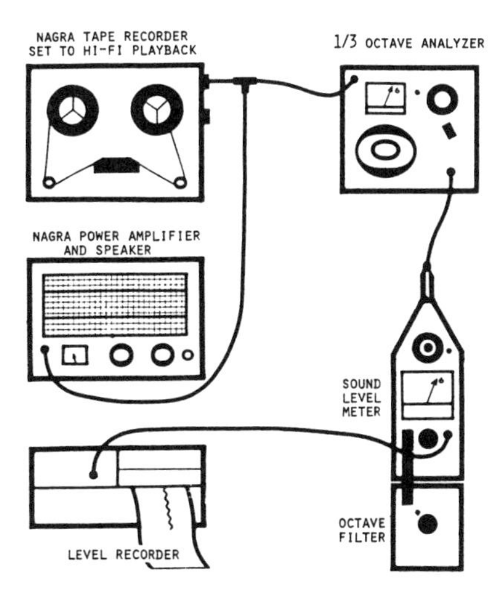

FIGURE 6. Details of laboratory arrangement of equipment for analyzing sound pressure levels of recorded "pink" noise and bird vocalizations.

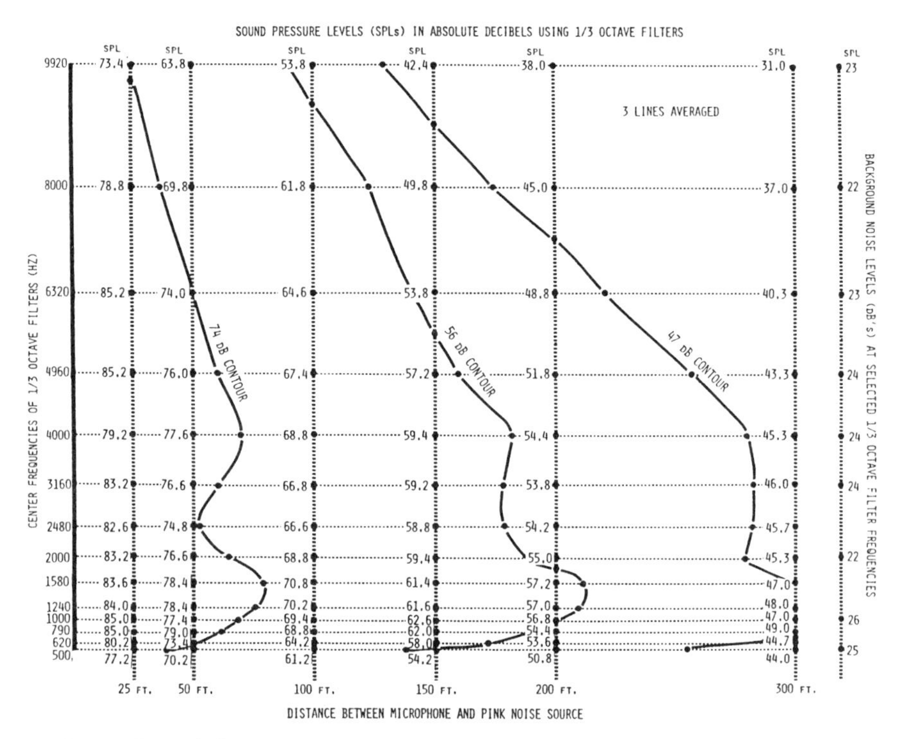

FIGURE 7. Partially completed sound-transmission isopleth for the coastat zone of Isla Santa Cruz, showing method of constructing isodecibel contour lines by connecting points on the grid of equal sound-pressure level (dB). Center frequencies of the one-third octave band filter are shown on the left vertical axis, and the distances between the sound source and six transect microphone positions are shown along the horizontal axis. Compare with isopleth (Fig. 8) where the contour interval is 3 dB.

spaced, sound is strongly attenuated; where they are widely spaced, sound is not strongly attenuated by the environment. If the slope of the contour line is more or less vertical across a bandwidth, it indicates that there is no frequency dependent disparity in sound attenuation at that distance from the sound source, and we may speak of an *unbiased* "acoustical aperture" whose width is delimited by the upper and lower frequencies in the bandwidth exhibiting such a uniform transmission efficiency. (The term "acoustical aperture," here coined for the phenomenon just described, should not be confused with the term "sound window" of Morton [1970, 1975] or "ground-effect window" of Marten [1980].) For example, in the partially completed sound transmission isopleth (Fig. 7), the acoustical aperture has a bandwidth of 2 kHz between 2 and 4 kHz over a

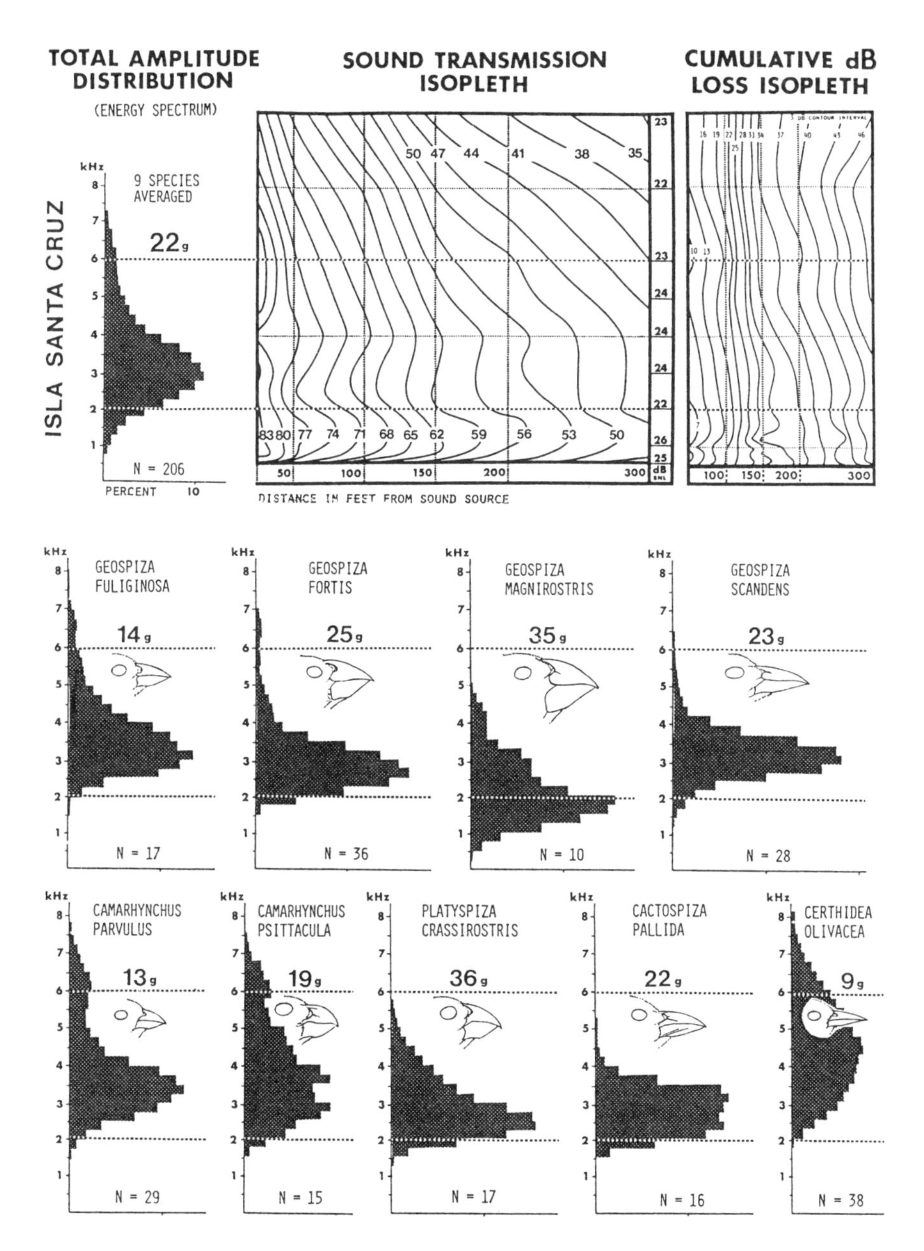

FIGURE 8. Song amplitude distributions of nine sympatric species of Galápagos finches on Isla Santa Cruz. Top row: Average of all species' distributions (left) and sound transmission isopleth for the inner coastal zone vegetation (right). Middle and bottom rows: Amplitude distributions of song populations of nine sympatric species.

distance of 300 feet from the sound source. We may arbitrarily delineate the width of the acoustical aperture by specifying how much disparity in amplitude may be allowed between highest and lowest frequencies in a song, beyond which signal integrity, and presumably auditory intelligibility are diminished or lost, *i.e.*, the result of a frequency-*biased* acoustical aperture. Since islands of the Galápagos differ in their floristic makeup, relative abundance and growth-form of species (Bowman, 1961; Wiggins & Porter, 1981; Hamann, 1979, 1981), it is not surprising that we do find persistent differences in frequency-dependent attenuation of sound from habitat to habitat and island to island.

For small-sized finches with comparatively small territories, the vocal signal need maintain its amplitude/frequency integrity over a relatively short transmission range, within which the acoustical aperture is usually quite broad. For example, in the sound isopleth for the inner coastal zone of Isla Santa Cruz (Fig. 8), it can be seen that at a distance of 50 feet from the sound source, the aperture has a frequency spread of about 4 kHz, *i.e.,* a bandwidth extending from about 2 to 6 kHz, with less than a 3-dB disparity between the upper and lower limits of the aperture. This relatively small difference in sound intensity, presumably, is insufficient to render the signal unintelligible. The width of the acoustical aperture is determined by distance from the sound source of the signal and the characteristics of the sound transmission pathway as primarily affected by vegetation. With increasing distance from the sound source, the bandwidth of the aperture generally narrows.

In all Galápagos environments examined so far, the amplitude disparity between specific high and low frequencies is invariably of lesser magnitude, the shorter the transmission distance, and of greater magnitude the longer the transmission distance. This condition probably explains why smaller birds, with smaller territories may have wider song bandwidths, utilizing the broad-band acoustical apertures, whereas larger birds with larger territories may have narrower song bandwidths, utilizing more constricted acoustical apertures available to them largely in the lower frequency ranges. Thus, a particular vocalization is considered to be well adapted to the transmission characteristics of its acoustical environment if the upper and lower frequency limits of the song bandwidth show no great disparity in amplitude loss over a distance equivalent to about the mean width of the species' breeding territory.

RESULTS AND DISCUSSION

Song frequencies. — Data presented in Fig. 9, support the conclusion of Greenewalt (1986) that there is poor correlation between the size of a songbird and the highest frequency of its song. For example, Isla Wolf *Geospiza magnirostris* (the second largest species of Galápagos finch by weight) and Isla Genovesa *Geospiza difficilis* (the second smallest species) sing some of the highest frequencies in their advertising songs, whereas Isla Española *Geospiza conirostris* (the third largest species) and Isla Santa Cruz *Certhidea olivacea* (the smallest species) sing some of the lowest frequencies in their songs. Furthermore, the so-called "whistle" call that is used in pair bonding and nest invitation by all species begins in the frequency range of 13-14 kHz (see Bowman, 1983).

Konishi (1970a) has suggested the possibility that the lowest frequencies of song may be correlated with the size of the syringeal internal tympaniform membrane. A comparison of data in Figs. 9 and 10 shows that such a correlation is none too good for Galápagos finches. Rather, the correlation seems to be much better when modal amplitudes of song frequency spectra are compared with average membrane dimensions. Although the smallest finch, *Certhidea olivacea* is capable of membrane vibrations at a fundamental frequency as low as in *Geospiza conirostris,* the third largest finch according to body weight, the absolutely smaller size of the membrane in *Certhidea,* presumably, causes it to vibrate at a modal frequency that is considerably higher and which is inherently "natural" for a species of its particular body size. But adaptive radiation has resulted in an array of species and insular populations with different body sizes (Figs. 9 and 10) with concomitant differentiation of "modal" frequencies as a secondary event, and not without direct adaptive value in specific environments. Evidence reported below clearly indicates that birds have the capacity, as a result of natural selection, to adapt their vocalizations to ecological conditions, often producing modalities in the energy spectra that are rather different from the modalities of the sensitivity spectra of their ears (*cf.* Dooling, *et al.*, 1971; Konishi, 1970a). Such fundamental disparities might be ameliorated in part by a source adjustment in the dB level of appropriate song frequencies.

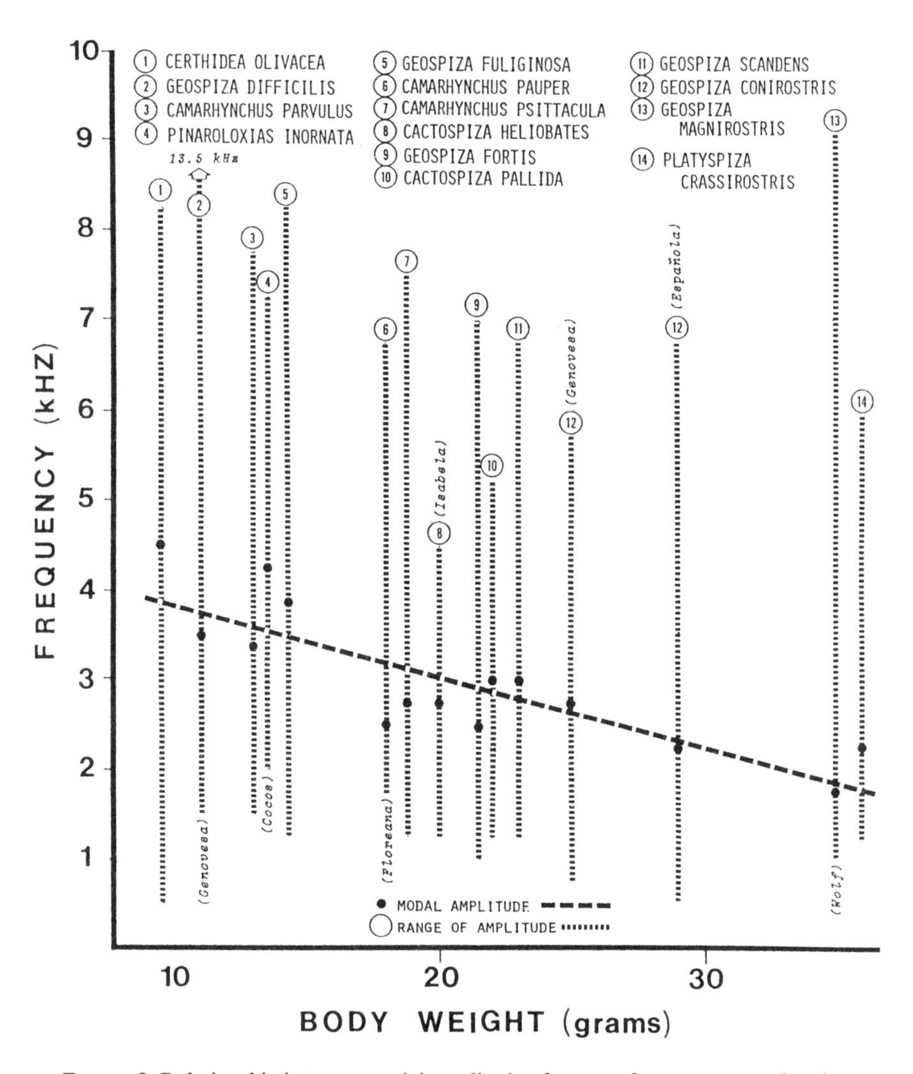

FIGURE 9. Relationship between modal amplitude of song to frequency spread and mean body weight for 14 species of Darwin's finches (*i.e.*, 13 species of Galápagos finches and 1 species of the Cocos Island finch). Data for males only are from Isla Santa Cruz birds, except as indicated.

Song amplitude in Galápagos finches appears to be correlated with absolute body size, with larger birds singing more loudly than smaller birds (Fig. 11 and Bowman, 1983, Figs. 122-126 and Table 18).

Song dialects. — Vocal dialects are said to occur when the songs or calls of conspecific birds in an area are very similar, but different from those of conspecifics in other areas. In other words, "dialects"

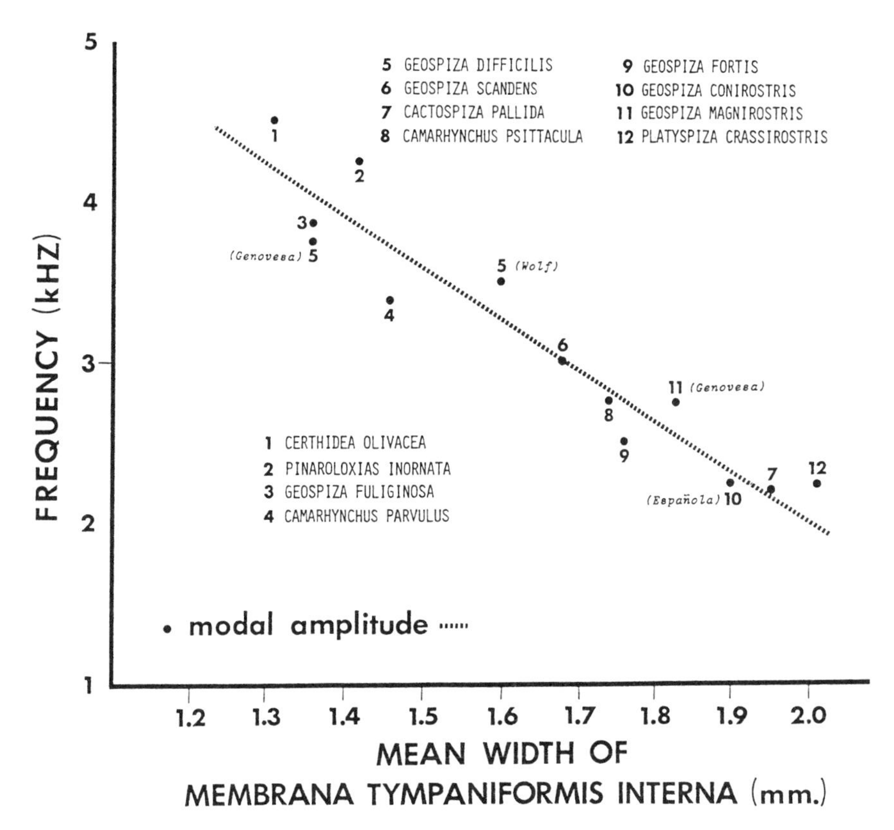

FIGURE 10. Relationship of modal amplitude of song to frequency and mean width of the *Membrana tympaniformis interna* in 12 species of Galápagos finches. Data for males only are from Isla Santa Cruz birds, except as indicated. Membrane data are from males in singing condition (Cutler, 1970).

are defined by the greater local occurrence of a particular song theme, which may occur less frequently or not at all elsewhere (Nottebohm, 1972).

Inter-island Song Dialects of Geospiza conirostris. — On Islas Española and Genovesa the song populations of *Geospiza conirostris* differ in many ways, the most obvious of which is frequency spread, *i.e.*, bandwidth (Fig. 2). Considering only the song bandwidths containing approximately 80-85% of the sound energy (shaded areas on sound transmission isopleths) we see that the frequency spread is 2.5 kHz (range 1.5-4.0 kHz) on Española and 1.0 kHz (range 2.0-3.0 kHz) on Genovesa. The frequency of the modal amplitude of song is 2.25 kHz and 2.75 kHz for Islas Española and Genovesa, respec-

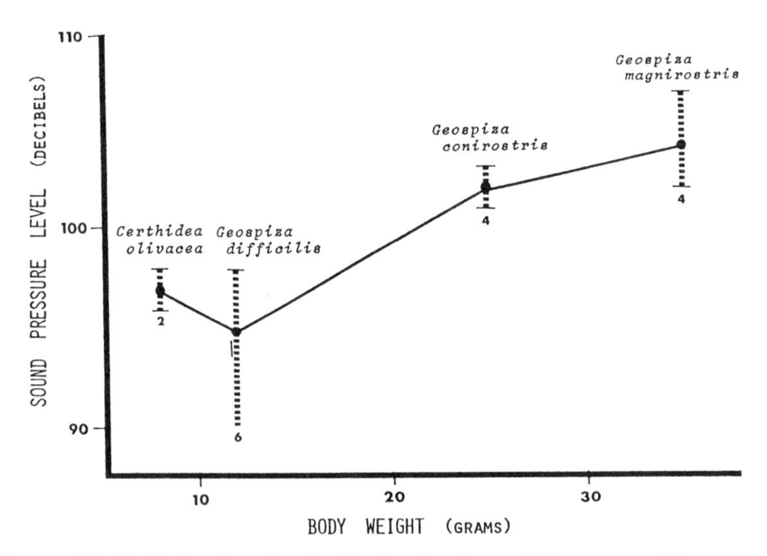

FIGURE 11. Decibel values for the "loudest" regions of the songs of four species of Galápagos finches on Isla Genovesa, arranged according to body weight. Solid lines connect mean values, and the broken vertical lines indicate range of variation in dB levels, with the number of songs analyzed shown at lower end of each line. (Compare with Table 3.)

tively. The broader frequency spread of the Española song appears to correlate with the broad unbiased acoustical aperture, as evidenced by the more or less vertical orientation and widely spaced isodecibel contour lines between approximately 1.5 and 4.0 kHz (Fig. 2). The sound transmission isopleth for Isla Genovesa suggests that the most efficient transmission bandwidth available to *G. conirostris* occurs between approximately 2 and 3 kHz, with very little latitude for variation into higher frequencies without developing a significant bias in the acoustical aperture, *i.e.*, a differential loss of amplitude between high and low frequencies in the bandwidth of such a magnitude as to endanger signal intelligibility.

With reference to the sound transmission isopleths, the difference in decay rates between highest and lowest frequencies of Española *G. conirostris* song over a distance of 300 feet from the sound source, is only one decibel (*i.e.*, 12 minus 11 dB, Table 1, line 1, columns F and G), and the difference in absolute intensity at 300 feet between highest and lowest frequencies in the central band which carries 80-85% of the song energy, is only two decibels (*i.e.*, 54 minus 52 d8; see Table 2, line 1, column D).

Comparable calculations for Genovesa *G. conirostris* yield a decay rate differential of only three decibels over 300 feet (Table 1, line 1, columns F and G) and an absolute intensity difference between highest and lowest frequencies of four decibels at the 300-foot distance (Table 2. line 3, column D).

These data suggest that long-distance sound transmission on Isla Genovesa is somewhat less efficient than on Isla Española, irrespective of frequency. A comparison of the isopleths for the two islands (Fig. 2) reveals a closer spacing of the isodecibel contour lines on Isla Genovesa, a condition indicating a relatively higher rate of sound attenuation. Whereas on Isla Genovesa the 3 kHz sound, for example, is reduced to an intensity of 44 dB at 300 feet from the sound source, on Isla Española the same frequency is reduced to only about 52 dB, thereby affecting an energy "saving" of 8 dB by the Española environment.

Perhaps a more revealing demonstration of the relative adaptiveness of the two *conirostris* song populations to their respective (concordant) environments can be made by projecting song dialects into discordant song environments, as is done in Table 1 and 2, lines 2 and 4. When the "narrow bandwidth" song of Genovesa *conirostris* is transmitted through the "wideband"-efficient discordant Española environment, there develops only a very small disparity in decay rates (*i.e.*, 1 dB; Table 1, line 2, column G) and absolute sound transmission intensities (*i.e.*, 1 dB; Table 2, line 2, column D) between high and low frequencies over a 300-foot transect. However, when the wide bandwidth" song of Española *conirostris* is transmitted through the "narrowband"-efficient discordant environment on Isla Genovesa, there develops a much larger disparity in decay rates (*i.e.*, 6 dB; Table 1, line 4, column G) and absolute sound intensities (*i.e.*,10 dB; Table 2, line 4, column D) between high and low frequencies over a 300-foot transect.

Although nothing is known about the sensitivity of the geospizine ear, some information is available (*cf.* Dooling, *et al.*, 1971; Konishi, 1970a) for songbirds and particularly for a distantly related "finch," the canary (*Serinus canarius*) which data we may apply, perhaps simplistically, to the condition in Galápagos finches (Table 2). I am assuming, for want of information to the contrary, that a source adjustment has not already been made in the songs of the birds in order to compensate for the differential frequency sensitivity of their ears. Consequently, it is necessary to subtract relative dB values (as derived from a graph by Dooling, *et al.*, 1971) from our absolute

TABLE 1. Comparison of attenuation rates of song frequencies of *Geospiza conirostris* on islas Española and Genovesa, and of *Geospiza scandens* on Isla Santa Cruz, Galápagos. Data for *G. scandens* are included for comparison because this species is the closest ecological equivalent of *G. conirostris* on other islands.

Column	A	B	C	D	E			F	G
Line	Song Environment	Song Frequency Class	Song Frequency Range Hz	Total Bandwidth of song (kHz)	Cumulative Loss in Amplitude (dB) Between Monitoring Stations 25/100′	100/200′	200/300′	Mean Rate of dB Loss per 100 ft. Between 25′ and 300′ Stations	Difference Between Mean HI-LO Rates of dB Loss
1.	Española song in Española environment	HI	6 500	6.00	14	11	8	11*	1
		LO	500		17	13	6	12	
2.	Genovesa song in Española environment	HI	5 750	4.25	14	11	8	11	1
		LO	1 500		10	14	7	10	
3.	Genovesa song in Genovesa environment	HI	5 750	4.25	17	17	11	15	3
		LO	1 500		10	13	12	12	
4.	Española song in Genovesa environment	HI	6 500	6.00	27	17	11	18	6
		LO	500		18	10	7	12	
5.	Santa Cruz song in Santa Cruz environment	HI	6 500	5.25	20	16	9	15	3
		LO	1 250		14	13	9	12	

* Mean Rate $= \frac{14 + 11 + 8}{3} = \frac{33}{3} = 11$

TABLE 2. Comparison of "intensity" differences between high and low song frequencies of *Geospiza conirostris* (Islas Española and Genovesa) and *Geospiza scandens* (Isla Santa Cruz) after compensation for relative differences in auditory sensitivities. Data for *G. scandens* are included for comparison because this this species is the closest ecological equivalent of *G. conirostris* on other islands.

Column	A	B	C	D				E	F				G			
Line	Song Environment	Song Frequency Class	Frequency (Hz) Range Containing 80—85 % of Song Energy (dB) Around Mode	Absolute Intensity (dB) of HI and LO Frequencies at Distances from the Sound Source of				Auditory Sensitivity Factor (Rel. dB)*	Absolute intensity Less Auditory Sensitivity Factor (Columns D minus E) at Distances of				Difference in dB Between Corrected HI and LO Frequencies at Distances from the Sound Source of			
				50′	100′	200′	300′		50′	100′	200′	300′**	50′	100′	200′	300′
1.	Española song	HI	3 750	74	67	59	52	6.0	68	61	53	46	1	1	5	5
	in Española environment	LO	1 500	80	75	61	54	13.2	67	62	48	41				
2.	Genovesa song	HI	3 000	74	69	59	51	1.0	73	68	58	50	4	9	3	3
	in Española environment	LO	2 250	74	64	60	52	5.0	69	59	55	47				
3.	Genovesa song	HI	3 000	74	64	53	43	1.0	73	63	52	43	1	0	2	1
	in Genovesa environment	LO	2 250	77	68	55	47	5.0	72	63	50	42				
4.	Española song	HI	3 750	74	61	48	40	6.0	68	55	42	34	0	7	7	7
	in Genovesa environment	LO	1 500	81	75	62	50	13.2	68	62	49	37				
5.	Santa Cruz song in	HI	3 750	78	69	54	45	6.0	72	63	48	39	0	1	3	4
	Santa Cruz environment	LO	2 500	75	67	54	46	3.2	72	64	51	43				

* Auditory sensitivity factors for *Serinus canarius* are from DOOLING, MULLIGAN & MILLER, 1971: 703, Fig. 3. On the average this species has its greatest sensitivity between 2.0 and 4.0 kHz. Sensitivity declines about 15 dB/octave as frequency is decreased below 2.0 kHz, declines about 25 dB between 4.0 and 8.0 kHz, and declines 13 dB between 8.0 and 9.0 kHz.

** All values are 10 dB or more above background noise levels.

sound intensity values (Table 2, column F). Presumably now the dB levels of the high and low frequency components of the songs at stations 50, 100, 200 and 300 feet better reflect the relative intensities as the birds might hear them in nature.

What emerges from an analysis of these data is a better appreciation and understanding of the adaptiveness of song to specific environments. Dialects in concordant environments (Table 2, lines 1 and 3, column G) are efficient in the sense that most of the song energy is concentrated in a transmission bandwidth that travels with the least distortion of all frequency components over great distance. Dialects in discordant environments (Table 2, lines 2 and 4, column G) develop notable intensity disparities (7-9 dB) between frequency components during short-range transmission (*e.g.*, 100 ft.), which, I assume, results in an early (*i.e.*, short-range) loss of intelligibility.

Differences in song ecology of the two island populations of *Geospiza conirostris* suggest the possibility that certain vagrant males might experience formidable difficulties in establishing themselves on another island, *e.g.*, Española to Genovesa transplant, because of possible mis-matching of song structure and transmission pattern, with all the attendant problems (*e.g.*, mate finding and territorial defense) normally mediated by song, in addition to unfamiliar food resources and behavioral adaptations for their full exploitation during times of resource scarcity. For example, an Española song of narrow bandwidth (such as R-73/10, Fig. 2) would be less maladapted than a song of wide bandwidth (such as R-46/2, Fig. 2) in the narrow bandwidth, energy-efficient channel of the Genovesa environment.

Field observations and laboratory experimentation indicate that geospizine song is culturally transmitted from father-to-offspring during the male parent-dependent fledgeling stage (Bowman, 1983; Millington & Price, 1985; Gibbs, 1990). Such filial tradition in song acquisition does, however, allow for some structural alteration of the signal such as occasionally takes place when a young orphan becomes misimprinted on the song of a foster male parent (Bowman, 1983).

PARALLEL EVOLUTION OF SONG STRUCTURE

(1) *Sympatric species in the coastal vegetation of Isla Santa Cruz.* — I attribute the widespread phenomenon of parallel structural variation in songs among sympatric species of finch, pictured in Figs. 12

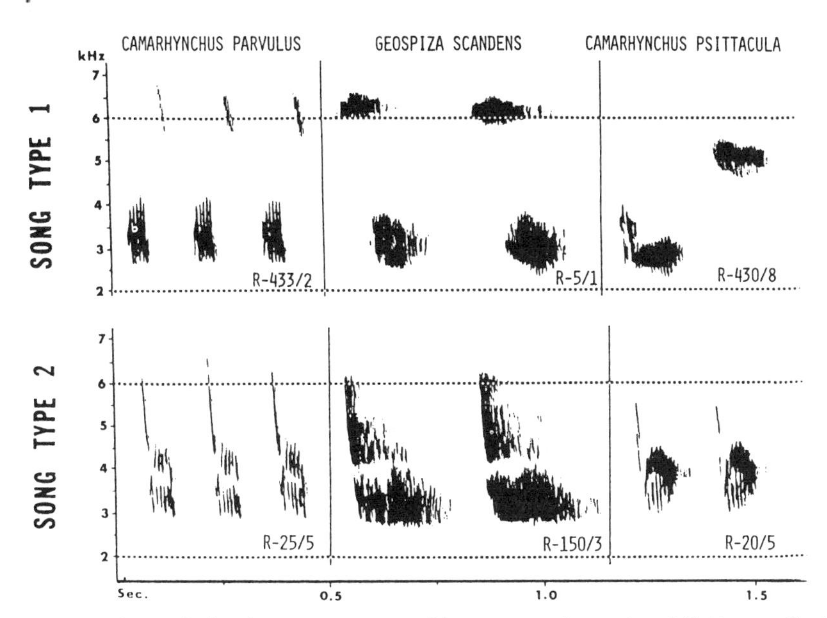

FIGURE 12. Parallelism in two song types of three sympatric species of Galápagos finches recorded at Academy Bay, Isla Santa Cruz. *Camarhynchus parvulus*, *Camarhychus psittacula,* and *Geospiza scandens* are different forms of "treefinches."

and 13, to similar acoustical "constraints" of the sound transmission environment and to similar behavior. Species whose songs are shown in Fig. 12 are essentially "tree-finches," ecologically; those shown in Fig. 13 are essentially "ground-finches," ecologically. Of course, since all available evidence indicates that the subfamily Geospizinae, to which all species of "Darwin's finches" belong, constitutes a monophyletic group (Bowman, 1961, 1983; Jo, 1983; Lack, 1945; Polans, 1983; and Yang & Patton, 1981), it is reasonable to assume that close genetic affinity accounts for some of the sharing of song parameters such as temporal subdivisions of their "basic" songs (see Bowman, 1983), but it probably does not account for much of the parallel structural patterns and energy distributions among finch species living in ecologically similar "sound environments."

Song bandwidths of long-distance song signals seem not to be correlated with a high or a low incidence of sympatry of the finches, but rather to be functionally related more to similar environmental features (*e.g.*, vegetation) affecting sound transmission.

(2) *Finch faunas of Islas Genovesa and Wolf.* — These northern islands support very similar finch faunas. Both have breeding populations of *Geospiza difficilis, G. magnirostris,* and *Certhidea olivacea*

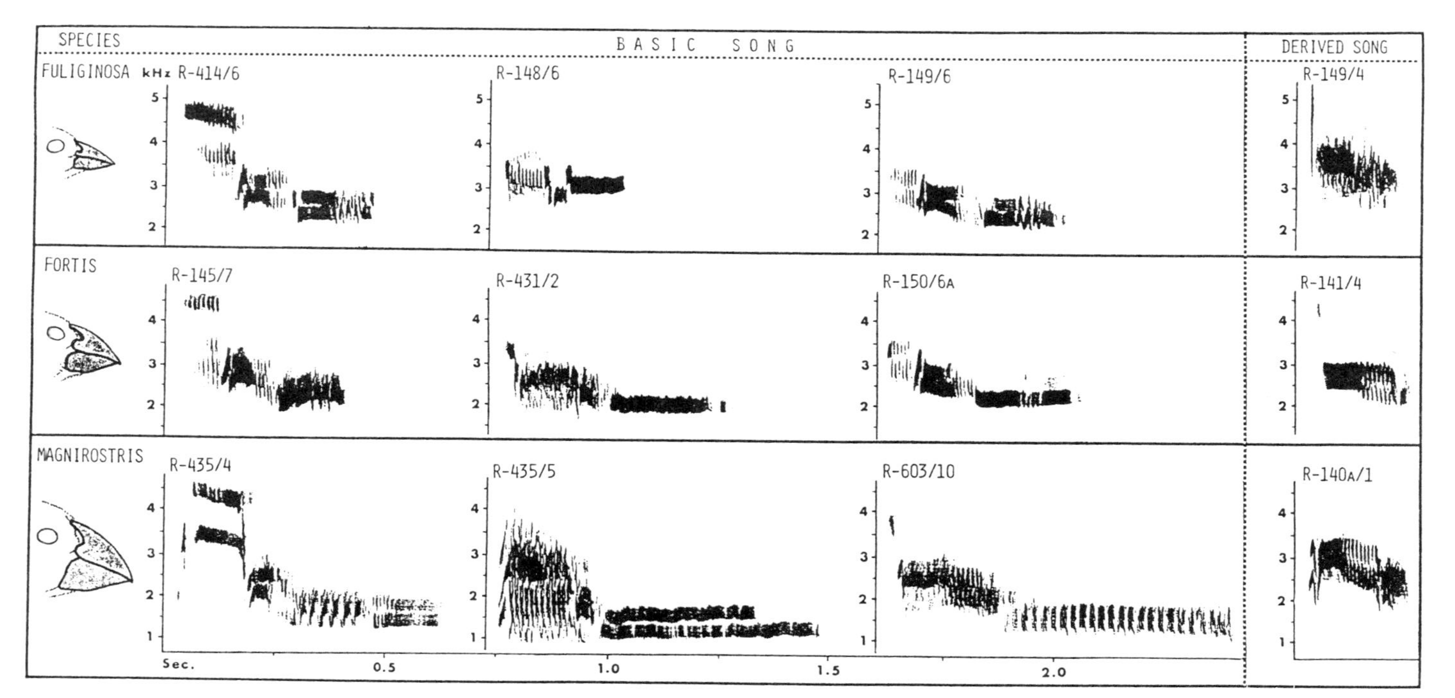

FIGURE 13. Parallelism in songs of three sympatric species of *Geospiza* recorded at Academy Bay, Isla Santa Cruz. Ecologically, *G. magnirostris*, *G. fortis* and *G. fuliginosa* are "ground-finches." Basic and derived song designations refer to a developmental scheme described by Bowman (1983).

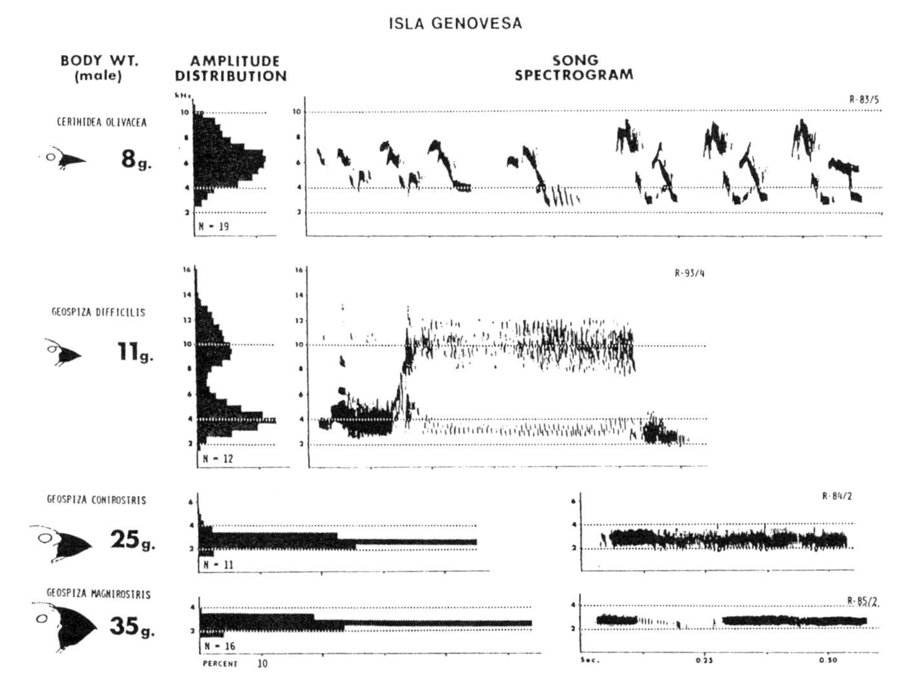

FIGURE 14. Comparison of song spectrograms, song amplitude distributions, and body size (wt.) of four species of Darwin's finches from Isla Genovesa, Galápagos. Body weight data courtesy of P.R.Grant.

(*cf.* Figs. 14 and 15). In addition to these species, Isla Genovesa has *Geospiza conirostris.* In 1965, Curio & Kramer reported this species breeding on Isla Wolf, but it has not since been recorded from this island.

The conspecifics on these two islands display marked song dialects (*cf.* Figs. 14 and 15). A comparison of amplitude distributions of the songs shows that the largest species, *G. magnirostris*, has a monomodal distribution of energy at 2.5 kHz on Genovesa, but a trimodal distribution on Wolf at 2.5, 5.0 and 8.0 kHz. A similar upward shift in frequency emphasis prevails in the case of *Geospiza difficilis*, with a bimodal distribution of energy on Genovesa at 4.0 and 10.0 kHz and a trimodal distribution on Wolf at 4.0, 9.5 and 14.0 kHz! The songs of the smallest species, *Certhidea olivacea,* although quite alike in their frequency distribution ranges on the two islands, differ in their amplitude distributions, with the Wolf population strongly skewed toward the higher frequencies (Fig. 15). It thus appears that there is a significant amplitude shift into the higher frequencies in all

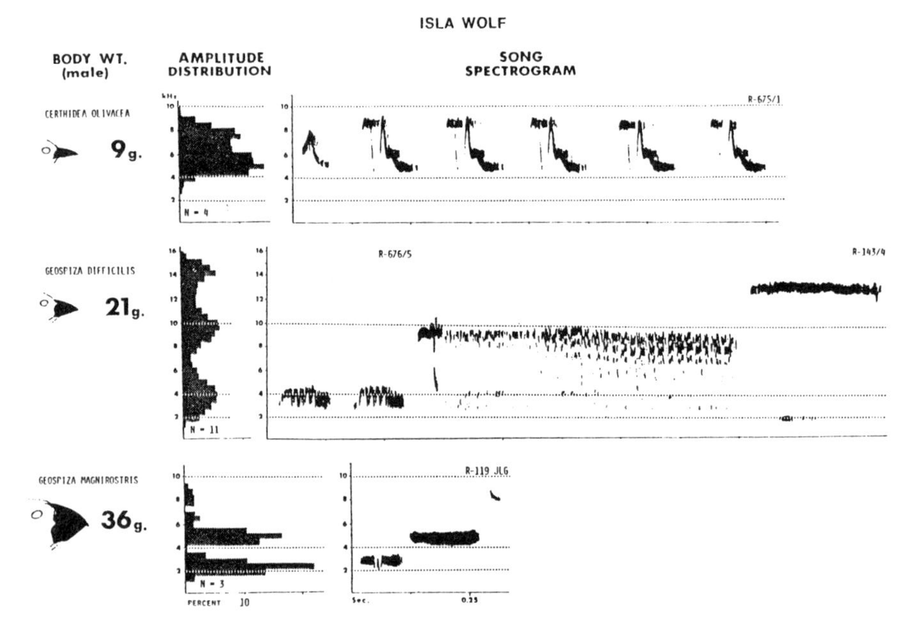

FIGURE 15. Comparison of song spectrograms, song amplitude distributions, and body size (wt.) of three species of Darwin's finches from Isla Wolf, Galápagos. Body weight data courtesy of P.R.Grant.

the songs of Wolf Island species, when compared with the situation in the Genovesa Island species.

If we combine the song amplitude distributions of all finch species for each island, we obtain histograms as shown in Fig. 16. With reference to these we can better appreciate how the total energy of the songs differs on Islas Genovesa and Wolf, and how this difference is related to the patterns of sound transmission. On Genovesa, the total energy distribution, although broad, is rather narrowly focused at the 2.5 kHz mode, whereas on Wolf the equally broad energy distribution is focused in the 2.0-2.5 kHz region, with a slight emphasis in the 8.0 kH region.

In relating song amplitude distributions to sound transmission isopleths, it should be noted that the isodecibel contour lines closest to the sound source on Wolf are almost vertically oriented for the first 50 feet of the transect. (Each contour line on the isopleth borders at a 3 dB change in sound pressure level.) In other words, any frequency within this sound spectrum (2.5-10.0 kHz) should experience little differential frequency-dependent loss of sound pressure in travelling 50 feet from the sound source. For frequencies between 2.5 and 10.0

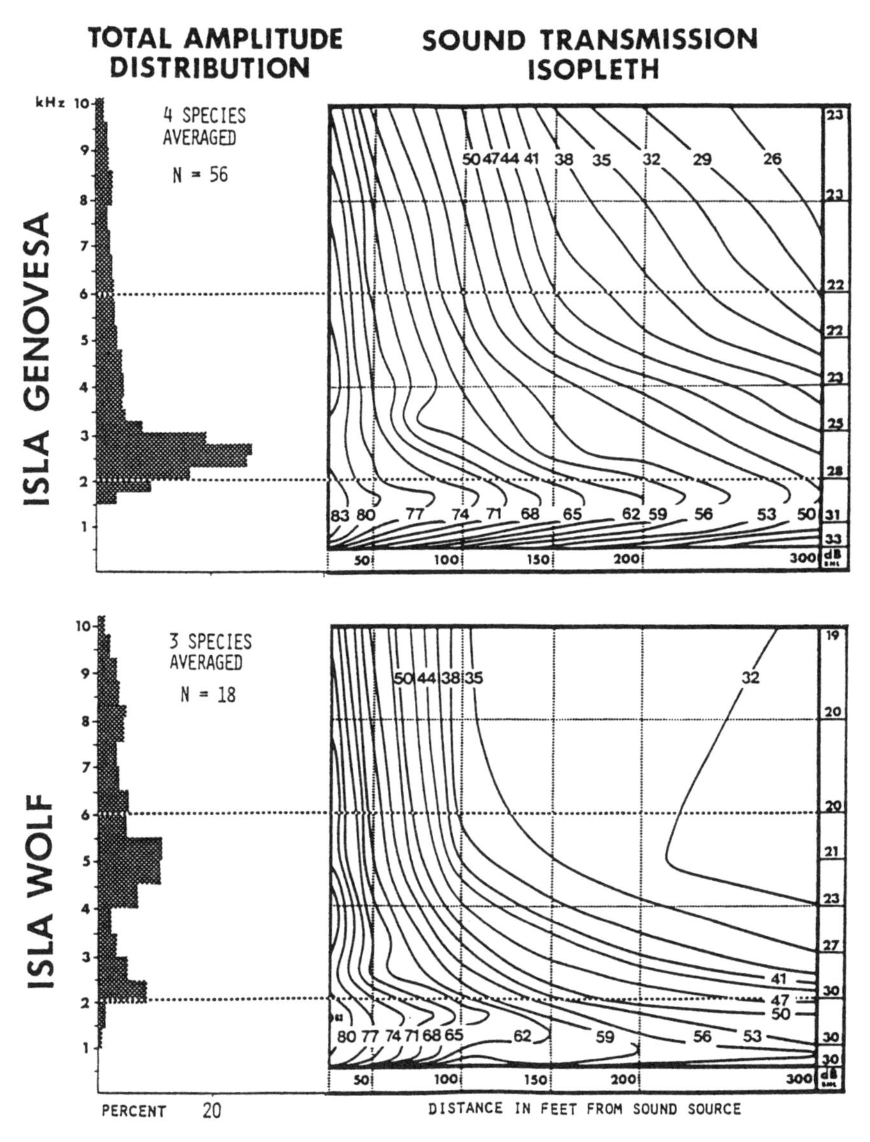

FIGURE 16. Average song amplitude distributions of all species of Darwin's finches on islas Genovesa and Wolf (left), and sound transmission isopleths (right). Background noise levels are shown at the right side of sound transmission isopleths.

kHz the average cumulative attenuation over this remarkably short distance is 15 dB (range 12-17 dB). For frequencies between 1.0 and 2.0 kHz, the average dB loss for the first 50 feet is 6 dB (range is 5-7 dB). So rapidly is sound attenuated by the Isla Wolf environment that at 4 kHz there is a 36 dB loss between the sound source and a point

100 feet distant! Compared with all other island studies, the Isla Wolf environment attenuates sound most rapidly, and this is true of all frequencies between 2 and 10 kHz. On Isla Wolf an "optimal" transmission frequency is achieved at a narrow frequency bandwidth centered at 2.0-2.5 kHz, at which an 18 dB loss is experienced at 100 feet from the sound source. Despite a slightly improved transmission at this frequency and a small modal peak on the amplitude histogram of combined songs, the bulk of the song energies is concentrated at the higher frequencies (Fig. 16).

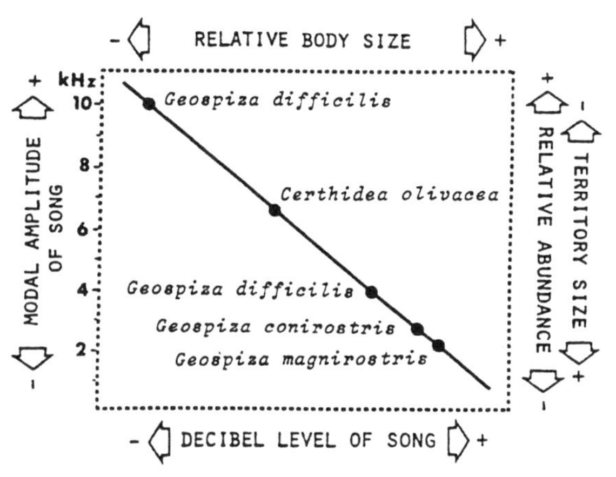

FIGURE 17. Generalized scheme showing relationships between frequency and decibel level of song, and relative body size, abundance, and territory size, as it pertains to four species of Darwin's finches on Isla Genovesa. Modified from Bowman (1979). *Geospiza difficilis* appears twice on the slope because of a pronounced bimodal amplitude distribution, namely 10 and 4kHz. The latter entry most closely reflects the relationship to weight and dB level of song; the former entry most closely reflects the relationships to territory size and relative abundance. Character states for each factor are indicated by plus and minus signs, i.e. + = high or large; - =low or small. Data on dB levels from Fig. 11. Territory sizes were estimated from Grant & Grant, 1980, Fig. 5. Note: *Geospiza difficilis* on Isla Genovesa is now best considered to be a form of *Geospiza fuliginosa* (Vagvolgyi & Vagvolgyi, 1989).

The comparatively poor long-distance sound transmission that characterizes the Isla Wolf environment has shaped the vocalizations of its song-bird inhabitants so as to "free" them from the environmental constraints which, as on Isla Genovesa, have kept them from emphasizing high frequencies in their songs. Indeed, even the large-sized *G. magnirostris* whose conspecifics on other islands of the Galápagos have songs with unimodal amplitude emphasis, for example, at 1.5-2.5 kHz for Isla Santa Cruz (Fig. 8), find remarkable "release" on Wolf (Fig. 15).

Field observations indicate that Isla Wolf has a rather dense and often patchy vegetation dominated by *Opuntia helleri* and *Croton scouleri* shrubs (Fig. 4e) which supports a high density population of finches, mainly *G. difficilis*. The abundance of finches and their foods, I speculate, has encouraged relatively small territory sizes, the consequence of which is as follows.

Territorial song, which is generally perceived as a long-distance communication signal, is operating at close range on Isla Wolf, and within an acoustical environment where there is no selective energetic disadvantage in singing songs punctuated with high frequencies. By "invading" a higher frequency spectrum the birds may be taking better advantage of increased information that can come from binaural localization, *i.e.*, directional hearing (Marler, 1955; Konishi, 1970b, 1973), especially where visibility is obfuscated by vegetation. The fact that *G. magnirostris* sings high-pitched sounds on Isla Wolf, despite the species' structural disposition to sing lower modal frequencies — because of its relatively large body size (Fig. 9) and concomitant internal tympaniform membrane (Fig. 10) — it may be presumed that a strong selective pressure exists favoring high frequency vocalizations. We probably can rule out any effect attributable to sound "masking" by coastal surf or sea-bird colonies, because the background noise level is about average for Galápagos environments (see dB levels at right side of isopleths, Fig. 16). If we compare the average song amplitude distribution of all finch species on Isla Genovesa (Fig. 16, top) with the auditory sensitivity curves presented by Dooling, Mulligan & Miller (1971) we find unimodal coincidence near 3 kHz. The same is not true for the averaged finch songs on Isla Wolf. Because of an amplitude shift to higher frequencies (Fig. 16, bottom), one might conclude, possibly erroneously, that the Wolf birds have different hearing sensitivities than their conspecifics on nearby Isla Genovesa. Whatever advantage might be conferred on the finches through intensive use of high frequencies in their songs must surely be founded in the unusual ecological (acoustical) conditions prevailing on Wolf Island.

Although we currently lack much data on territory size for Galápagos finches on most of the islands, it is here assumed that larger bodied species such as *G. magnirostris* and *G. conirostris* hold larger territories on Isla Genovesa than do the smaller bodied species such as *G. difficilis* and *Certhidea olivacea*. This notion is in agreement with the general findings of territorial studies made on continents (Welty & Baptista, 1988), notwithstanding the many complicating circumstances that might alter such a relationship in non-songbird species. In addition, if we assume that a more or less direct relationship exists between population density and territory size, then we may infer from the relative abundance data on species of *Geospiza* obtained by Abbott, Abbott & Grant (1977) on Isla Genovesa, that *G. difficilis* (the smallest species of the genus) has a smaller territory size than *G. magnirostris* (the largest species), with *G. conirostris* (the medium-sized species) having a territory somewhat intermediate in size between the two former species, but probably closer to that of *G. magnirostris*.

Defense of a large area is facilitated by the use of a long-distance vocal signal. The development of a low-frequency modal amplitude (3 kHz) and a narrow bandwidth in the songs of the larger species on Isla Genovesa, has prompted a highly adapted advertising song that is capable of exploiting the most energy efficient transmission channel in the environment (*cf.* Figs. 14 and 16). The second smallest species on Genovesa, *G. difficilis*, shows a bimodal amplitude distribution in its song and a significantly wider bandwidth that is broadly spread into the high frequencies, insignificantly "occupied" by the songs of the larger species. The greater relative abundance and presumed smaller territory sizes in *G. difficilis* and *Certhidea olivacea*, have lessened the "need," so to speak, for a far-flung vocal signal, and may have encouraged the use of higher frequencies whose energies, when compared with those of lower frequencies, are not disproportionately squandered by the environemnt or made functionally ineffective within the shorter communication distances that are likely to be involved. It may be in error to always equate operationally, at least, "territorial song" to "long-distance signal." In general, smaller-sized and more abundant species of song birds, in possession of smaller territories, probably produce lower amplitude songs, spread over a broader frequency spectrum, than is the case with larger-sized and less abundant species of song birds. This results in a reduction of effective transmission distance of songs, which, nevertheless, meets the requirements necessary, among other things, for territory maintenance and mate attraction (*cf.* Heuwinkel, 1978).

Data on sound pressure levels of territorial songs of free-living Galápagos finches are available for Islas Genovesa and Wolf (Table

TABLE 3. Sound pressure levels of advertising songs of four species of Darwin's finches. (The asterisk * refers to points in time along the length of the song as shown in Bowman, 1983. Decibel values are adjusted so as to represent sound pressure levels at approximately one inch from the beak of the birds.)

Species	Island	Mean body weight g	No. of songs	Mean dB level by song region* 1	2	3	4	5
Geospiza magnirostris	Genovesa	35	4	91	93	104	103	—
Geospiza conirostris	Genovesa	25	4	100	100	102	100	101
Geospiza difficilis	Genovesa	11	6	73	74	95	94	78
Geospiza difficilis	Wolf	21	1	104	92	94	92	—
Certhidea olivacea	Genovesa	8	2	82	93	91	97	—

3). These indicate that larger species have somewhat more powerful (louder) songs than smaller species. For example, on Isla Genovesa the two larger species, *G. magnirostris* and *G. conirostris*, have some intensities ranging from 91 to 104 dB, and from 100 to 102 dB, respectively, whereas the two smaller species, *G. difficilis* and *Certhidea olivacea*, have ranges from 73 to 93 dB and from 84 to 99 dB, respectively. If we examine only the song regions with the highest dB values (Fig. 11), we see that the larger species inject more energy into their songs than the smaller species.

A picture emerges of the relationship between decibel level and modal frequency of song, on the one hand, and relative body and territory sizes, and relative abundance on the other hand (Fig. 18; *cf.* Jilka & Leisler, 1974).

(3) *Intra-populational variation in songs of* Camarhynchus parvulus *on Isla Santa Cruz.* — On the south-facing slope of Isla Santa Cruz, the parid-like, small-beaked tree finch, *Camarhynchus parvulus,* breeds from the coastal cactus plains up through the fog-drip *Scalesia pedunculata* forest at an elevation of about 700 feet (see Figs. 4 a and 4 b, and Bowman, 1983). Although the birds exhibit no obvious morphologi-

cal differentiation throughout this range, their galaxy of song shows a clinal shift along a lowlands-highlands transect. Songs heard in the arid coastal zone are structurally more diverse than those heard in the highlands zone (Fig. 18). Whereas the *Scalesia* forest highlands are "home" for basically one form of song, namely type "B," the lowlands play host to many song forms whose extreme is type "A." The difference is striking. Coastal song types are reminiscent of certain vocalizations of the plain titmouse (*Parus inornatus*), the dark-eyed junco (*Junco hyemalis*) and the chipping sparrow (*Spizella passerina*) of western North America, whereas the highlands song shows some acoustical resemblance to the territorial "chick-a-dee" utterances of the black-capped chickadee (*Parus atricapillus*) and the plain titmouse of North America. This writer has never heard an "A" type song in the *Scalesia* forest, but the "B" type song is sometimes heard in the lowlands.

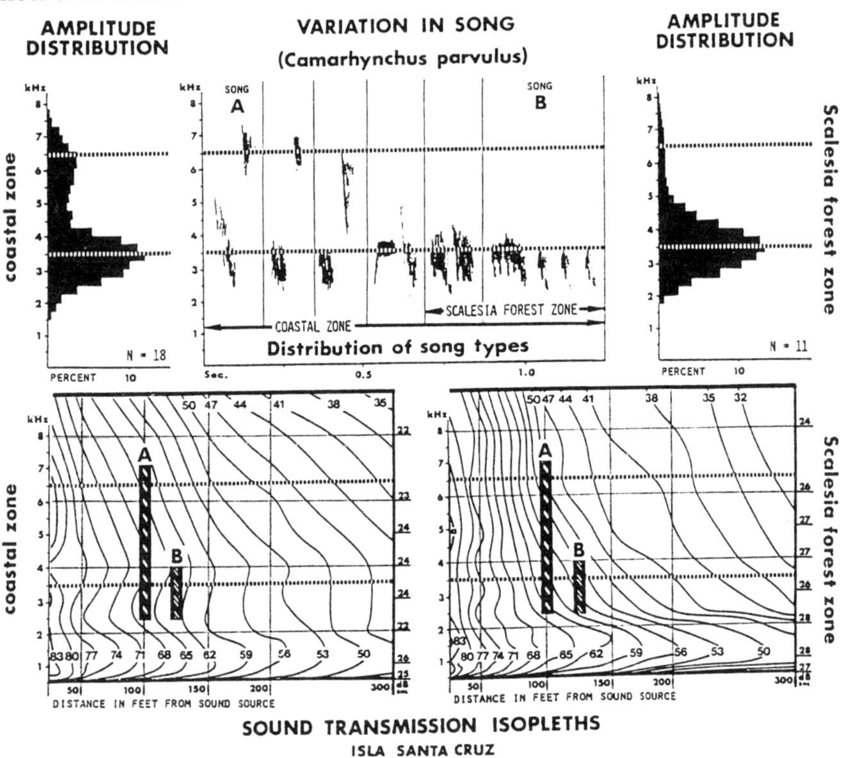

FIGURE 18. Examples of song types of *Camarhynchus parvulus* recorded on the south-facing slope of Isla Santa Cruz (above), showing histograms of amplitude distribution of populations in the coastal zone (left) and *Scalesia* forest zone (right). Vertical bars on isopleths (below) represent song types "A" and "B." See text for explanation.

In order to explain this rather unusual song distribution (one which could change with the continued destruction of the *Scalesia* forest for purposes of agriculture), let us compare amplitude distributions for highland and lowland songs with their respective sound transmission isopleths. To facilitate the analysis, and to better appreciate the adaptive significance of the distribution of the song types, two black vertical bars, labelled "A" and "B" of different lengths and cross-hatchings are superimposed on the lowland and highland isopleths at points 100 ft. and 125 ft. distances, respectively, from the sound source. The bar length corresponds approximately to the frequency spread of the respective song types. The coastal zone type "A" song is high-pitched to the human ear and has a very broad frequency spread (2.5-7.0 kHz) and a total bandwidth of 4.5 kHz. The *Scalesia* forest highlands song type "B" has a comparatively narrows frequency spread (2.5-4.0 kHz) and a total bandwidth of 1.5 kHz, and consequently sounds lower-pitched. In Panamá, Morton (1970) noted a similar difference in pitch between the songs of birds living in "tropical" forests and those living in more open habitat.

The song type "B" travels with minimal distortion for hundreds of feet in the coastal zone vegetation, largely because of the narrow bandwidth, which allows this signal to "fit" into the 2.5-4.0 kHz transmission channel where the contour lines of the coastal zone isopleth (Fig. 18, lower left) proceed like square-fronted wavefronts through the thorn-cactus "forest" (Fig. 4a). Sound pressure levels of high and low frequencies decrease uniformly over the entire bandwidth and transmission field (see Table 4, line 3, columns E and F). Clearly, this vocal signal is well adapted for long-range communication in the coastal zone of Isla Santa Cruz. The same can be said about song type "B" in the *Scalesia* forest zone, except that here the overall rate of attenuation is somewhat greater in the first 150 feet from the sound source as evidenced by the more closely positioned and vertically oriented contour lines (Fig. 18, lower left). With increasing distance, the type "B" sound of the highlands becomes ever so slightly less efficient than in the coastal zone. Because the rate of dB loss for the type "B" song frequencies is higher in the *Scalesia* forest zone than in the coastal zone (compare cumulative dB loss figures in Table 4, lines 3 and 4, column E), the song intensity, at least in the higher frequencies, is reduced to a level that is only slightly above the background noise level at a distance between 200 and 300 feet from

TABLE 4. Comparison of "intensity" differences between high and low frequencies in song types "A" and "B" of *Camarhynchus parvulus* in the coastal and *Scalesia* forest zones of Isla Santa Cruz.

Column	A	B	C	D				E				F			
Line	Song Environment	Song Frequency Class	Maximum Frequency Range (Hz)	Absolute Intensity (dB) of HI and LO Frequencies at Distances from the Sound Source of				Cumulative dB loss at HI and LO Frequencies at Distances from the Sound Source of				Difference in Absolute Intensity (dB) between HI and LO Frequencies (Column D) at Distances from Sound Source of			
				50′	100′	200′	300′	50′	100′	200′	300′	50′	100′	200′	300′
1.	Song "A" in coastal zone environment*	HI	7 000	74	65	49	40	11	20	36	44	1	2	5	6
		LO	2 500	75	67	54	46	8	16	29	27				
2.	Song "A" in *Scalesia* forest environment	HI	7 000	74	50	40	[31]*	8	32	42	51	0	13	6	7
		LO	2 500	74	63	46	38	9	20	37	45				
3.	Song "B" in coastal zone environment	HI	4 000	78	69	54	45	1	10	25	34	3	2	0	1
		LO	2 500	75	67	54	46	8	16	29	37				
4.	Song "B" in *Scalesia* forest environment	HI	4 000	73	59	43	[34]	10	24	40	49	1	4	3	4
		LO	2 500	74	63	46	38	9	20	37	45				

* Brackets around an intensity value indicate a level of 10 dB or less above background noise level and, presuuably, close to the auditory threshold.

the sound source (Table 4, line 4, column D), even though the highest frequencies of song type "B" are relatively low when compared with those of song type "A." At this distance we must assume that the signal loses intelligibility and is soon lost totally in the environment. These findings suggest the possibility that *Camarhynchus parvulus* may have a somewhat smaller territory in the moister, resource-richer *Scalesia* forest zone than in the drier coastal zone, where food resources for this species are less abundant.

Song type "A" has a strong emphasis on the higher frequencies (see Fig. 18) which decay at a slightly higher rate than the lower frequencies, resulting in an every increasing disparity between the two from 1 dB to 6 dB at 50 and 300 feet, respectively, from the sound source (See Table 4, Line 1, Column F).

Thus the wideband "A" song type seems to be almost as well adapted as the narrow-band "B" type song to the first 100 feet of the coastal transmission field on Isla Santa Cruz, but with the "B" type song, with its predominantly lower frequencies, slightly more efficient at greater distances (up to 300 feet). See Table 4, lines 1 and 3, column F.

If type "A" song were sung in the highlands of Santa Cruz Island it would be non-adaptive beyond the 50-100 ft. range because of the serious frequency-dependent amplitude disparity in transmission rates (*i.e.*, 13 dB) that sets in and causes early disruption of the information content of the song (see Table 4, line 2, column F).

CONCLUSIONS

In view of the numerous inter- and intra-island differences (and similarities) in the pattern of sound transmission through vegetation in Galápagos environments, it was probably inevitable that highly variable patterns in the songs of the finches should have evolved through the process of natural selection. The cultural, *i.e.*, non-genetic, evolution of dialects (ecophenotypes) is essentially a tracking of the local environments by differential preservation of populations whose vocal signals are designed for optimal transmission of identity and motivational information. Mutual occupancy and exploitation of Galápagos environments for species maintenance has fostered parallel ecological adjustments in the advertising songs of the finches. These adjustments take the form of modifications of the communication signal that tend to minimize transmission loss by matching song structure with specific sound propagation characteristics of the environment. Genetic differentiation of populations has probably been facilitated by energy economizing adaptations of song to specific environments, thereby tending to promote philopatry.

ACKNOWLEDGMENTS

The author wishes to thank the following institutions for their financial support of this research: National Science Foundation, National Geographic Society, American Museum of Natural History, New York Zoological Society, California Academy of Sciences, EARTHWATCH, and San Francisco State University. Portions of this report have been adapted from previous papers by Bowman (1979, 1983) in which full acknowledgement of individual help is given.

Literature Cited

Abbott, I, L. K. Abbott, and P. R. Grant. 1977. Comparative ecology of Galápagos ground finches (*Geospiza* Gould): Evaluation of the importance of floristic diversity and interspecific competition. *Ecol. Monogr.* 47:151-184.

Beebe, W. 1924. *Galápagos: World's End.* G. P. Putnam's Sons, New York, N.Y.

Beebe, W. 1926. *The Arcturus Adventure.* G. P. Putnam's Sons, New York, N,Y.

Bowman, R. I. 1961. Morphological differentiation and adaptation in the Galápagos finches. *Univ. California Publ. Zool.* 58:1-302.

Bowman, R. I. 1979. Adaptive morphology of song dialects in Darwin's finches. *Jour. Ornithol.* 120:353-389.

Bowman, R. I. 1983. The evolution of song in Darwin's finches. Pages 237-537 *in* R. I.Bowman, M. Berson, & A. E. Leviton, eds., *Patterns of Evolution in Galápagos Organisms.* American Association for the Advancement of Science, Pacific Division, San Francisco, California.

Chappuis, C. 1971. Un example de l'influence du milieu sur les émission vocales des oiseaux: l'évolution des chants en forêt équitoriale. *Terre et Vie* 25:183-202.

Curio, E. and P. Kramer. 1964. Vom Mangrovefinken (*Cactospiza heliobates* Snodgrass and Heller). *Zeitschr. Tierpsychol.* 21:223-234.

Cutler, B. D. 1970. Anatomical studies on the syrinx of Darwin's finches. *M.A. Thesis.* San Francisco State University, San Francisco, Calif.

Dooling, R. J., J. A. Mulligan, and J. D. Miller. 1971. Auditory sensitivity and song spectrum of the common canary (*Serinus canarius*). *Jour. Acoust. Soc. America* 50:700-709.

Gibbs, H. L. 1990. Cultural evolution of male song types in Darwin's medium ground finches, *Geospiza fortis. Anim. Behav.* 39:253-263.

Gifford, W. E. 1919. Field notes on the land birds of the Galápagos Islands and of Cocos Island, Costa Rica. *Proc. California Acad. Sci.* 2:189-258.

Grant, P. R. and B. R. Grant, 1980. The breeding and feeding characteristics of Darwin's finches on Isla Genovesa, Galápagos. *Ecol. Monogr.* 50:391-410.

Hamann, O. 1979. On climatic conditions, vegetation types, and leaf size in the Galápagos Islands. *Biotropica* 11:101-122.

Hamann, O. 1981. Plant communities of the Galápagos Islands. *Dansk Bot. Arkiv* 34:13-163.

Heuwinkel, H. 1978. Der Gesang des Teichrohsängers (*Acrocephalus scirpaceus)* unter besonder Berücksichtigung der Schalldruckpegel ("Lautstärke") *Verhältnisse Jahrb. Ornithol.* 119:425-461.

Hunter, M. L. and J. R. Krebs. 1979. Geographical variation in the song of the great tit (*Parus major*) in relation to ecological factors. *Jour. Anim. Ecol.* 48:759-785.

Jilka, A., and B. Leisler. 1974. Die Einpassung dreier Rohrsängerarten (*Acrocephalus schoenobaenus, A. scirpaceus, A. arundinaceus*) in ihre Lebensräume in Bezug auf das Frequenzspektrum ihrer Reviergesänge. *Jahrb. Ornithol.* 115:192-212.

Jo, N. 1983. Karyotypic analysis of Darwin's finches. Pages 201-217 *in* R. I. Bowman, M. Berson, & A. E. Leviton, eds., *Patterns of Evolution in Galápagos Organisms.* American Association for the Advancement of Science, Pacific Division, San Francisco, Calif.

Konishi, M. 1970a. Comparative neurophysiological studies of hearing and vocalizations in songbirds. *Zeitschr. vergl. Physiol.* 66:257-272.

Konishi, M. 1970b. Evolution of design features in the coding of species-specificity. *American Zool.* 10:67-72.

Konishi, M. 1973. Locatable and nonlocatable acoustic signals for barn owls. *American Nat.* 107:778-785.

Lack, D. 1945. The Galápagos finches (Geospizinae): A study in variation. *Occ. Pap. California Acad. Sci.* (21):1-152.

Marler, P. 1955. Characteristics of some animal calls. *Nature* 176:6-7.

Marten, K. L. 1980. Ecological sources of natural selection on animal vocalizations, with special reference to the African wild dog (*Lycaon pictus*). *Ph.D. Thesis*. University of California, Berkeley, Calif.

Marten, K., and P. Marler. 1977. Sound transmission and its significance for animal vocalization. I. Temperate habitats. *Behav. Ecol. Sociobiol.* 2:271-290.

Marten, K., D. Quine, and P. Marler. 1977. Sound transmission and its significance for animal vocalization. II. Tropical forest habitats. *Behav. Ecol. Sociobiol.* 2:291-302.

Mayr, E. 1963. *Animal Species and Evolution.* Harvard Univ. Press, Cambridge, Mass.

Millington, S. J. and T. D. Price. 1985. Song inheritance and mating patterns in Darwin's finches. *Auk* 102:342-346.

Morton, E. S. 1970. Ecological sources of selection on avian sounds. *Ph.D. Thesis.* Yale University, New Haven, Conn.

Morton, E. S. 1975. Ecological sources of selection on avian sounds. *American Nat.* 109:17-34.

Mundinger, P. C. 1982. Microgeographic and macrogeographic variation in acquired vocalizations of birds. Page 147-208 *in* D. F. Kroodsma, E. H. Miller, & H. Oullet, eds., *Acoustic Communication in Birds*, vol. 2. Academic Press, New York, N.Y.

Polans, N. 0. 1983. Enzyme polymorphisms in Galápagos finches. Pages 219-236 *in* R. I. Bowman, M. Berson, & A. E. Leviton, eds., *Patterns of Evolution in Galápagos Organisms.* American Association for the Advancement of Science, Pacific Division, San Francisco, Calif.

Rothschild, W. and E. Hartert. 1902. Further notes on the fauna of the Galápagos Islands. *Nov. Zool.* 9:373-418.

Ryan, M. J. and B. K. Sullivan. 1989. Transmission effects on the temporal structure of the advertisement call of two species of toads, *Bufo woodhousii* and *Bufo valliceps*. *Ethology* 80:182-185.

Ryan, M. J. and W. Wilczynski. 1991. Evolution of intraspecific variation in the advertisement call of a cricket frog (*Acris crepitans,* Hylidae). *Biol. Jour. Linnean Soc.* 44:249-271.

Snodgrass, R. E. and E. Heller. 1904. Papers from the Hopkins-Stanford Galápagos Expedition, 1898-1899. XVI. Birds. *Proc. Washington Acad. Sci.* 5:231-372.

Welty, J. C. and L. Baptista, 1988. *The Life of Birds,* 4th ed. Saunders College Publishing, New York.

Vagvolgyi, J. and M. W. Vagvolgyi. 1989. The taxonomic status of the Small Ground-Finch, *Geospiza* (Aves:Emberizidae) of Genovesa Island, Galápagos, and its relevance to interspecific competition. *Auk* 106:144-148.

Wiggins, I. L., and D. M. Porter. 1971. *Flora of the Galápagos Islands.* Stanford University Press, Stanford, Calif.

Wiley, R. H., and D. G. Richards. 1978. Physical constraints on acoustic communication in the atmosphere: Implications for the evolution of animal vocalizations. *Behav. Ecol. Sociobiol.* 3: 69-94.

Yang, S. Y. and J. L. Patton. 1981. Gene variability and differentiation of the Galápagos finches. *Auk* 98:230-242.

ABSTRACTS OF PAPERS PRESENTED AT THE SYMPOSIUM HELD ON THE OCCASION OF THE 100TH ANNIVERSARY OF THE BIRTH OF GÖTE TURESSON

Genecology and Plant Taxonomy. KENTON L. CHAMBERS, Department of Botany and Plant Pathology, Oregon State University, Corvallis, OR 97331.

Gote Turesson introduced his genecological categories for population classification at an opportune time for plant taxonomy. The 1920's saw the dramatic impact of genetic concepts and experimental methods on evolutionary biology. The idea that the variation patterns observed by taxonomists had, first of all, an adaptive genetic basis, and secondly, were largely determined by measurable environmental factors, greatly stimulated experimental research and helped keep taxonomy in the mainstream. Attempts by various workers to apply the categories of ecotype, ecospecies, etc. directly in formal (i.e., Linnean) classifications were seldom successful. Debates about clinal variation, kinds and degrees of genetic isolation among populations, and the effects of differing breeding systems on morphological and genetic variation helped to advance taxonomy conceptually and in everyday practice. Disagreements still exist as to whether formal classification should be mainly morphological and "general purpose," or should define taxa by "phylogenetic" criteria. A synthetic species concept is possible, which places genetic isolation in a proper context along with the ecologic and demographic features of populations.

Adaptive Patterns in Populus trichocarpa

Joan Dunlap, College of Forest Resources AR-10, University of Washington, Seattle, WA 98195.

A common-garden study of Populus trichocarpa T.& G. was initiated in 1985 when clonal material was collected from two mesic (Hoh, Nisqually) and two xeric (Dungeness, Yakima) river valleys. In 1986, the material was planted in two replicate plantations, one in a maritime climate (Puyallup, WA) and the other in a continental one (Wenatchee, WA). Two-year results showed greater survival and growth at Puyallup. Stem volumes at that site decreased significantly in this order: Nisqually > Hoh > Dungeness > Yakima. Mesic-origin trees were also larger in other traits, e.g., leaf size and branch length, than xeric-origin trees. Although spring flush generally did not vary among rivers, Yakima trees had the earliest autumn budset and leaf fall, and the greatest susceptibility to Melampsora leaf rust. There were marked differences between elevational groups in Nisqually and Yakima trees, respectively. Greater stem volume, earlier spring flush, and later budset in lower Nisqually trees relative to upper ones were associated with a longer growing season in the lower reaches. For the Yakima, climate associated with physiography and elevation influenced the timing of spring flush. Lower stem volume, earlier budset, and earlier leaf loss of lower Yakima trees were mainly due to their infection with Melampsora leaf rust; the patterns may also reflect an adaptive response to greater moisture stress in the lower Yakima. The transition between mesic and xeric ecotypes coincides with a sharp change in atmospheric moisture along the Yakima.

Ecotypic variation in response to serpentine soils.

ARTHUR R. KRUCKEBERG (Department of Botany, University of Washington, Seattle, 98195)

Turesson's ecotype concept has changed over time: from ecotype to ecotypic variation to race or provenance to interspecific variation. Yet its essential truth persists - genotypic response to habitat. Response to the edaphic factor, especially to soils of highly contrasting attributes is a widespread phenomenon. It is typified by infraspecific differences in tolerance to serpentine soils (iron-magnesium rich soils often high in Ni, Co, Cr). Case histories of western North American species (annuals to woody taxa) illustrate the racial differentiation. Akin to the ecotypic response to serpentines are the genotypic responses to other heavy metal soils or mine tailings (with Zn, Pb, et al.). Beyond showing infraspecific tolerance to such substrates, are questions of physiological and genetic bases for the tolerance. For serpentines, further research on tolerance to low Ca, high Mg, and high Ni levels is needed. Ecotypic response to serpentines or to a heavy metal can be judged as an initial stage in the genesis of edaphic endemic species.

Ecotypic Variation in Marine Angiosperms: Seagrasses and Mangroves (CALVIN McMILLAN, Department of Botany, University of Texas, Austin, TX 78713)

The submerged seagrasses in *Thalassia*, *Halodule*, and *Syringodium* and the mangroves in *Avicennia*, *Rhizophora* and *Laguncularia* show ecotypic differentiation that is related to the latitudinal variation of their habitats in the Gulf of Mexico and the Caribbean. Populations of seagrasses and mangroves show physiological adaptation to the colder winter climates in the northern Gulf of Mexico and those in more tropical habitats show various degrees of physiological dysfunction under these chilling conditions. The latitudinal gradients in Atlantic habitats have resulted in numerous examples of ecotypic variation among marine angiosperms, but those in the Indo-Pacific region are less intensively studied.

Edaphic Races in Forest Trees. C.I. MILLAR; J.L. JENKINSON. Institute of Forest Genetics, USDA Forest Service, Pacific Southwest Research Station, Berkeley, CA 94701.

In forest trees, breeding systems favor abundant, long-distance gene flow, and large genetic neighborhoods. Patterns of variation are trait-specific, and depend on individual selection:gene-flow gradients. We demonstrate these effects in a review of edaphic differentiation in forest trees. Literature from three sets of contrasting soil types--wet vs. dry, ultramafic vs. fertile, and severe podsols ("pygmy-forest" soils) vs. fertile--is analysed regarding variation in isozyme, growth traits, and eco-physiological traits. Allelic differences in isozymes rarely correlate with soil of origin, and when they do, they may relate as much to historical patterns of population isolation as to soil gradients. Height-growth differences measured in common-garden tests generally indicate weak or no patterns associated with soil of origin when measured in seedlings. When transplant experiments are carried out for many years, strong correlations with soils of origins occasionally result from evaluations of older trees. By contrast, strong patterns relating to soil adaptation have been measured in early tests in root traits, nutrient uptake, and other physiological traits. We illustrate these situations with new data from tests of soil adaptation in Pinus ponderosa and Pinus muricata. In the former, trees from populations on serpentine and fertile soils were grown in common-gardens tests in the soils of origin. After 20 years, significant differences in height, stem volume, foliar mass, and nutrient levels resulted, although at one-year, differences had not been significant. In P. muricata, populations from severe podsols and fertile soils were analyzed for differences in isozymes, germination, phenology, foliar nutrient levels, growth and morphology in common-gardens, and root mass/root morphology in pot-tests. Distinct differences in morphology and one isoyzme locus appear related to

ENVIRONMENTAL AND GENETIC LIMITS TO ADAPTIVE POPULATION DIFFERENTIATION IN QUANTITATIVE TRAITS

Gerrit A.J. Platenkamp and Ruth G. Shaw
Department of Botany and Plant Sciences
University of California, Riverside

The environmental and genetic covariance structures of traits related to fitness set limits to the evolution of locally adapted populations. For quantitative traits a response to natural selection may be limited by (1) a lack of genetic variation, (2) a large environmental effect on the phenotype, (3) lack of genotype-environment interaction variance, and (4) adverse genetic correlations. The extent of these limits to natural selection can be investigated by the quantitative genetic analysis of fitness traits of families or clones that were reciprocally transplanted between field sites. Only recently multivariate statistical techniques have been developed and implemented that can be used to analyze the genetic and phenotypic structure of quantitative traits in this type of field experiment. We will use the example of recent population differentiation in the introduced grass, Anthoxanthum odoratum, in Northern California grasslands to illustrate these techniques. Populations show significant genetic differences. However, there does not appear to be much potential for future adaptive differentiation.

From the mountains to the prairies to the ocean white with foam: Papilio zelicaon makes itself at home.

ARTHUR M. SHAPIRO (Department of Zoology, University of California, Davis, CA 95616)

Papilio zelicaon, the Anise Swallowtail, has an extraordinary range, from sea level to the Sierra Nevada alpine zone, from the coastal fog belt to high desert. Its phenology varies immensely in tandem with that of its hosts. Univoltine races may occur in close geographic proximity to highly multivoltine ones, and there is circumstantial evidence for rapid evolution of seasonality in association with introduced hosts. Electrophoretic studies indicate little genomic reorganization has occurred, and suggest that substantial gene flow is still occurring among the various ecotypes.

Ecotypic variation in butterflies. MICHAEL C. SINGER, DANIEL A. VASCO, CHRISTIAN D. THOMAS, and RAYMOND R. WHITE (Department of Zoology, University of Texas, Austin, TX)

The checkerspot butterfly Euphydryas editha showed local adaptation both in life-history traits and in behavior patterns associated with host plant use. As in many plant species, there was genetically based reduction of size with increasing altitude. Suites of behavioral adaptations to particular host plant species also showed genetic variation among habitats. Samples of insects from habitats as little as 200 m apart differed significantly in host preference, but with some overlap. Large samples from populations 15 km apart failed to show any overlap of host preference. These examp;es illustrate the spatial scale on which ecotypic variation occurs in this species. In contrast, butterflies that are highly vagile show little geographical variation, and the variation that exists is not clearly related to local conditions.

Göte Turesson: The Pacific Coast connection.

RICHARD B. WALKER (Department of Botany, University of Washington, Seattle 98195)

Turesson was born in 1892 in Malmö, Sweden, and received his early education there. At age 20 he came to Spokane, Washington, to the home of an aunt. While working for a year at various jobs, and not yet educated as a botanist, he nonetheless started observations of the effects of slope and aspect on the distribution of Douglas-fir. The next year he enrolled at the University of Washington, Seattle, and came under the influence and tutelage of the three faculty members of the Department of Botany: Theodore C. Frye, morphologist; John W. Hotson, mycologist; and George B. Rigg, physiologist and ecologist. Along with earning the B.S. in 1914 and the M.S. in 1915, he finished and published the work on Douglas-fir (Bull. Torrey Bot. Club 41:337-345), reported with C. Thom a new species of Penicillium (Mycologia 7:281-287), and carried out a detailed ecological study of the western skunk cabbage (Amer. Jour. Bot. 3:189-209). Thus when he returned to Sweden in 1915, and matriculated at the University of Lund for doctoral studies, he was already an active botanist with three substantial publications to his credit. He came to the United States once again, in 1934, when invited to attend the annual meeting of the Carnegie Institution of Washington, held that year in Stanford, California.

GLOSSARY FOR ECOGEOGRAPHIC RACES

acoustical aperture: sound bandwidth delimited by upper and lower frequencies in which there is no frequency-dependent disparity

agamospecies: species reproducing asexually.

agrestal: weed in crops; weed of cultivated lands.

aneuploidy: state of having fewer or more chromosomes than an exact multiple of the haploid number.

apomixis: asexual reproduction, either vegetatively or by seeds from cells other than ovules (akin to parthenogenesis).

atrophied foretarsi: vestigial or reduced appendages (in insects).

autogamy: producing seeds by way of self-pollination.

cladistic analysis: a technique for determining the phylogeny of a group of organisms based on an analysis of the distribution of derived characters across a spectrum of taxa (characters described as apomorphic = uniquely derived character or character state found in one taxon only; synapomorphic = derived character or character state shared by two or more taxa).

cline (clinal, adj.): a character gradient; interpopulational variation in a single character, usually across heterogeneous habitat.

coenospecies: (1) a group of taxonomic species of common evolutionary origin (a Turesson term), usually equivalent to a section of a genus or an entire genus. (2) species so related that they may exchange genes among themselves to a limited extent through hybridization.

comimetic species: species that mimic each other.

comparium: all of the coenospecies (which see) between which hybridization is possible either directly or through intermediaries.

constraints (limits) on evolution: genetic, developmental or morphological attributes impeding evolutionary change.

deme: population; group of sexually reproducing individuals sharing in a common gene pool.

dioecism (dioecious, adj.): having sexes (unisexual flowers) on separate individual plants.

eclosion: in insects, hatching from an egg or from an imago (larval stage).

ecologic race: local population system within a species, adapted to a particular environment (similar to ecotype).

ecophenotype: a non-genetic modification of the phenotype in response to an environmental condition (Mayr, 1963).

ecotype: 1) a population or system of populations arising as a result of the genotypic response of a species to a particular habitat (term coined by Turesson). (2) all of the members of a species which are fitted to survive in a particular kind of environment within the total range of the species.

ecotypic differentiation: ability of a species to develop local populations genetically endowed for given habitats (contrasts with phenotypic plasticity and general purpose genotype).

electrophoretic: denotes a technique for distiguishing different allelic or genic versions (isozymes, allozymes) of a given enzyme.

epigamic behavior: devices to attract the opposite sex (*e.g.*, colors displayed in courtship).

epiphenomenon: a phenomenon which occurs with and seems to result from another.

fitness: the average contribution of one allele or genotype to the next (or succeeding) generation(s) compared with that of other allelles or genotypes.

genecology: the study of genetypically controlled responses of populations to habitats, and the resulting ecotypic differentation (which see).

general purpose genotype (gpg): wide adaptive tolerance of a single genotype to a range of environments (contrasts with ecotypic differentation).

genetic architecture: limits imposed by the genotype to genetic change; also, the make-up of the total genotype.

genetic constraints: limits imposed by the genotype to genetic change.

heritability: the degree to which a gene (or genotype) may be expressed phenotypically; *e.g.*, low vs. high heritability.

instar: an insect at a particular stage between moults.

introgression: interspecific hybridization followed by back-crossing of the F_1 to one or the other parent.

limits (constraints) to growth: structural or functional impediments to normal growth.

mimetic: relating to, or characterized by, or exhibiting mimicry.

monocarpic: flowering (and fruiting) followed by death of the individual plant (similar to annual life form).

oviposition: act of egg-laying in insects.

phenetics: study of phenotypic realization of genotypes.

phenotypic plasticity (plasticity): wide range of character expression (phenotypic response) of a given genotype.

pink noise: a sound spectrum characterized by a sound pressure level decrease of 3 decibels per octave (compare with "white noise," *q.v.*)

polymorphic loci: gene loci contributing to a quantitative trait, each such locus having a small phenotypic effect.

phytometer: using the performance of a living plant as a measure of a given environmental factor.

polyploidy (polyploids): multiples of chromosome sets above or below the diploid state (polyploids have such a chromosome status).

polytopic subspecies: similar geographic race occurring in more than one place.

r-K *continuum*: range of response between r-selection and K-selection; a given species will exhibit some mix of the two modes of selection. (r-selected species are those in unstable habitats with high reproductive potential; K-selected species are ones of low reproductive potential fitted for stable predictable habitats).

ruderal: weed of wayside or waste places, occurring in disturbed habitats other than cultivated fields.

typological species: classical concept of a species, allowing for little or no variation; species fitting the "type" gestalt.

univoltine vs. multivoltine: refers to insects having one (*vs.* more than one) brood per year.

weed: any plant (other than a crop plant) that thrives under human disturbance.

white noise: a sound spectrum characterized by a constant energy distribution per octave (compare with "pink noise," *q.v.*).

Index

Index

C

G

H

N

O

P

Q

R

S

PRODUCTION INFORMATION

Typesetting: Pacific Division AAAS, California Academy of Sciences, Golden Gate Park, San Francisco, California. Typeset in Adobe Systems 12pt. Times Roman using Ventura Publisher® Windows-version 4.1 on an ALR® 486/33 microcomputer and output on a LazerMaster™ 1000/4 Personal Typesetter laser printer.

Printing and Binding: Braun-Brumfield, Inc., Ann Arbor, Michigan/Richard Thunes, San Francisco, CA. The text is printed on 60# Natural Hi-Bulk acid-free paper that meets the guidelines for permanence and durability established by the Committee on Publications Guidelines and Book Longevity of the Council on Library Resources. The cover is Roxite C vellum that conforms to ANSI L29 specifications. Endpapers are acid-free, Rainbow® Felt, Ivory-FD by Ecological Fibers, inc.

Date of Publication: 10 July 1995.

Place of Publication: San Francisco, California, USA.

Number of Copies (First Printing): 300.